Safety, Reliability, Human Factors, and Human Error in Nuclear Power Plants

T0174253

Safety, Reliability, Human Factors, and Human Error in Nuclear Power Plants

B.S. Dhillon
Department of Mechanical Engineering
University of Ottawa
Ottawa, Canada

CRC Press
Taylor & Francis Group
Boca Raton London New York

CRC Press is an imprint of the
Taylor & Francis Group, an **informa** business

CRC Press
Taylor & Francis Group
6000 Broken Sound Parkway NW, Suite 300
Boca Raton, FL 33487-2742

First issued in paperback 2019

ISBN-13: 978-1-138-08099-7 (hbk)
ISBN-13: 978-0-367-89064-3 (pbk)

Visit the Taylor & Francis Web site at
http://www.taylorandfrancis.com

and the CRC Press Web site at
http://www.crcpress.com

This book is affectionately dedicated to my son-in-law Mandeep.

Contents

Preface

Each year billions of dollars are being spent in the area of nuclear power generation to design, manufacture, operate, and maintain various types of systems around the globe. In 2011, there were around 436 commercial nuclear reactors operating in the world that generated about 16% of the world's electricity.

Many times, the systems used in nuclear power plants fail due to safety, reliability, human factors, and human error-related problems. For example, during the period from 1990–1994, approximately 27% of commercial nuclear power plant outages in the United States occurred due to human error.

Although over the years a large number of journal and conference proceedings articles related to safety, reliability, human factors, and human error in nuclear power plants have appeared, there is no book that covers all four of these topics within its framework to the best of the author's knowledge. This causes a great deal of difficulty for information seekers because they have to consult many different and diverse sources.

Thus, the main objective of this book is to combine all these topics into a single volume in order to eliminate the need to consult many different and diverse sources to obtain the desired information. The sources of most of the material presented are listed at the end of each chapter in the reference section. These will be useful to readers if they wish to delve more deeply into a specific area or topic. The book contains a chapter on mathematical concepts and another chapter on introductory safety, reliability, human factors, and human error concepts, which are considered useful for understanding the content of subsequent chapters. Furthermore, two more chapters are devoted to methods considered useful for performing safety, reliability, human factor, and human error analyses concerning nuclear power plants.

The topics covered in the book are treated in such a manner that the reader will require no previous knowledge to understand the contents. At appropriate places, the book contains examples along with their solutions, and at the end of each chapter there are numerous problems to test the reader's comprehension in the area. An extensive list of publications dating from 1960–2015 (relating directly or indirectly to safety, reliability, human factors, and human error in nuclear power plants) is provided at the end of this book to give readers a view of the intensity of developments in the area.

The book is composed of 12 chapters. Chapter 1 presents the need for and the background concerning nuclear power plants; nuclear power plant safety, reliability, human factor, and human error-related facts, figures, and examples; important terms and definitions; useful sources for obtaining information on safety, reliability, human factors, and human error in nuclear power plants; and the scope of the book. Chapter 2 reviews mathematical concepts considered useful for understanding subsequent chapters. Some of the topics covered in the chapter are Boolean algebra laws, probability properties, probability distributions, and useful mathematical definitions.

Chapter 3 presents various introductory aspects of safety, reliability, human factors, and human error. Chapter 4 presents a number of methods considered useful for performing safety, reliability, human factors, and human error analyses in nuclear power plants. These methods include technique of operations review, root cause analysis, interface safety analysis, hazards and operability analysis, failure modes and effect analysis, man-machine systems analysis, fault tree analysis, the Markov method, and the probability tree method. Chapter 5 is devoted to human reliability analysis methods for nuclear power stations. Some of the methods covered in the chapter are standardized plant analysis risk–human reliability analysis, the cognitive reliability and error analysis method, the technique for human event analysis, the technique for human error rate prediction, and the Human Error Assessment and Reduction Technique.

Chapter 6 presents various important aspects of safety in nuclear power plants. Some of the topics covered in the chapter are nuclear power plant fundamental safety principles, nuclear power plant specific safety principles, management of safety in nuclear power plant design, safety-related requirements in the design of specific nuclear power plant systems, and deterministic safety analysis for nuclear power plants. Chapter 7 is devoted to nuclear power plant accidents and covers topics such as the Three Mile Island accident; the Chernobyl accident; the Fukushima accident; lessons learned from the Three Mile Island, Chernobyl, and Fukushima accidents; and comparisons of the Chernobyl and Three Mile Island accidents and of the Fukushima and Chernobyl accidents.

Chapter 8 presents various important aspects of reliability and maintenance programs for nuclear power plants. Some of the topics covered in the chapter are reliability program objectives and requirements; guidance for developing reliability programs; maintenance program objectives, scope, and background; and guidance for developing maintenance programs. Chapter 9 is devoted to human factors and human error in nuclear power generation and covers topics such as aging nuclear power plant human factor-related issues; human factor-related issues that can have a positive impact on decommissioning of nuclear power plants; human factors review guide for next-generation nuclear reactors; human error facts, figures, and examples concerning nuclear power generation; occurrences caused by operator errors during operation in commercial nuclear power plants; and causes of operator errors in commercial nuclear power plant operations.

Chapter 10 presents various important aspects of human factors and human error in nuclear power plant maintenance. Some of the topics covered in the chapter are the study of human factors in power plant maintenance; elements relating to human performance that can contribute to an effective maintenance program in nuclear power plants; useful human factors methods to assess and improve nuclear power plant maintainability; nuclear power plant maintenance error-related facts, figures, and examples; causes of human error in nuclear power plant maintenance; useful guidelines for human error reduction and prevention in nuclear power plant maintenance; and methods for performing maintenance error analysis in nuclear power plants.

Chapter 11 is devoted to human factors in nuclear power plant control systems and covers topics such as human performance-related advanced control room technology issues and control room design-related deficiencies that can lead to human error, human engineering discrepancies in control room visual displays, human factor-related evaluation of control room annunciators, and recommendations for overcoming problems when digital control room upgrades are undertaken without considering human factors in the design. Finally, Chapter 12 presents eight mathematical models for performing safety, reliability, and human error analysis in nuclear power plants.

The book will be useful to many individuals, including engineering professionals working in the area of nuclear power generation; engineering administrators; engineering undergraduate and graduate students; power system engineering researchers and instructors; safety, reliability, human factors, and psychology professionals; and design engineers and associated engineering professionals.

I am deeply indebted to many individuals, including family members, colleagues, friends, and students for their invisible input. The invisible contributions of my children are also appreciated. Last but not least, I thank my wife, Rosy, my other half and friend, for typing this entire book and for her timely help in proofreading.

B.S. Dhillon
Ottawa, Ontario, Canada

Author

B.S. Dhillon, PhD, is a Professor of Engineering Management in the Department of Mechanical Engineering at the University of Ottawa. He has served as the Chairman/Director of the Mechanical Engineering Department/Engineering Management Program for over 10 years at the same institution. He is the founder of the probability distribution named the Dhillon Distribution/Law/Model by statistical researchers in their publications around the world. He has published over 234 journal papers and 155 conference proceedings articles on reliability engineering, maintainability, safety, engineering management, and so on. He is or has been on the editorial boards of 12 international scientific journals. In addition, Dr. Dhillon has written 45 books on various aspects of health care, engineering management, design, reliability, safety, and quality published by Wiley (1981), Van Nostrand (1982), Butterworth (1983), Marcel Dekker (1984), and Pergamon (1986), to name a few. His books are being used in over 100 countries, and many of them have been translated into languages such as German, Russian, Chinese, and Persian (Iranian).

He has served as General Chairman of two international conferences on reliability and quality control held in Los Angeles and Paris in 1987. Dr. Dhillon has also served as a consultant to various organizations and bodies and has many years of experience in the industrial sector. He has lectured in over 50 countries, including keynote addresses at various international scientific conferences held in North America, Europe, Asia, and Africa. In March 2004, Dr. Dhillon was a distinguished speaker at the Conf./Workshop on Surgical Errors (sponsored by the White House Health and Safety Committee and the Pentagon), held on Capitol Hill in Washington, DC.

Dr. Dhillon attended the University of Wales where he received a BS in electrical and electronic engineering and an MS in mechanical engineering. He received a PhD in industrial engineering from the University of Windsor.

1

Introduction

1.1 Background

The history of nuclear power plants goes back to 1954, when the first five megawatt (MW) nuclear plant for generating electricity for commercial use started to operate at Obnisk, Russia [1]. In 1956, four 50 MW reactors started to operate at Calder, United Kingdom, and in 1957, a 60 MW pressurized light water reactor (PWR) began operation at Shippingport, United States [1]. Nowadays, as per Reference 2, nuclear power plants generate about 16% of electricity globally, and there are over 440 commercial nuclear reactors operating in 30 different countries, with another 65 reactors under construction.

Needless to say, the systems used in nuclear power plants often fail due to safety, reliability, human factors, and human error-related problems. For example, in the United States during the period from 1990–1994, approximately 27% of commercial nuclear power plant outages were due to human error [3]. Since 1960, a large number of publications directly or indirectly related to safety, reliability, human factors, and human error in nuclear power generation have appeared. A list of over 340 such publications is provided in the Appendix.

1.2 Safety, Reliability, Human Factors, and Human Error-Related Facts, Figures, and Examples

Some of the facts, figures, and examples directly or indirectly concerned with safety, reliability, human factors, and human error in nuclear power plants are as follows:

- A study of the commercial nuclear power plant outages in the United States during the years 1990–1994 reported that about 27% of the outages were due to human error [3].

- A study performed by the U.S. Nuclear Regulatory Commission (NRC) of licensee reports revealed that around 65% of the nuclear system failures were due to human error [4].
- As per References 5 and 6, approximately 70% of nuclear power plant operation-related errors appear to have a human factor-related origin.
- Approximately 54% of the incidents due to human errors in Japan during the years 1969–1986 led to automatic shutdowns of nuclear reactors, and around 15% of those resulted in power reduction [7].
- As per Reference 8, in nuclear power plants in 1985, the distribution of causes reflected a roughly 50/50 split between equipment-associated causes versus those attributable to human performance-related problems; in 1990, human performance-related problems increased to about 70% of all causes.
- A study of over 4400 maintenance-related history records for the period from 1992–1994 concerning a boiling water reactor (BWR) nuclear power plant reported that approximately 7.5% of all failure records could be clearly attributed to human errors directly or indirectly related to maintenance activities [9,10].
- As per Reference 11, in Japanese nuclear power plants during the period from 1965–1995, 199 human errors occurred, out of which 50 of them were concerned with maintenance-related tasks.
- During the period from February 1975 to April 1975, a study of 143 occurrences of operating U.S. commercial nuclear power plants revealed that approximately 20% of the occurrences were due to operator-related errors [12,13].
- As per References 14 and 15, in South Korean nuclear power plants during the years 1978–1992, 255 shutdowns occurred, out of which 77 of them were human induced.
- As per Reference 16, a study of major incident/accident reports of nuclear power plants in South Korea indicated that approximately 20% of the events occurred due to human error.
- In 1989 on Christmas Day in the state of Florida, two nuclear reactors were shut down due to maintenance-related errors and caused rolling blackouts [17].
- As per Reference 18, in the area of nuclear generation, in 1990, a study of 126 human error-related significant events revealed that approximately 42% of the problems were directly or indirectly linked to modification and maintenance.
- As per Reference 19, the Three Mile Island Nuclear Power Plant accident that occurred in 1979 in the United States was the result of human-related problems.

- As per Reference 20, in South Korean nuclear power plants, approximately 25% of unexpected shutdowns were due to human errors, out of which over 80% of human errors resulted from usual testing and maintenance-related tasks.
- As per References 19 and 21, the Chernobyl Nuclear Power Plant accident in Ukraine in 1986, widely considered as the worst accident in the history of nuclear power generation, was the result of human-related problems.

1.3 Terms and Definitions

This section presents some useful terms and definitions directly or indirectly related to safety, reliability, human factors, and human error in nuclear power plants [21–30].

- **Safety:** This is conservation of human life and the prevention of damage to items as per specified mission requirements.
- **Reliability:** This is the probability that an item/system will carry out its stated function satisfactorily for the desired period when used according to the specified conditions.
- **Human factors:** This is the study of the interrelationships between humans, the tools they utilize, and the surrounding environment in which they perform tasks and live.
- **Human error:** This is the failure to carry out a stated task (or the performance of a forbidden action) that could result in disruption of scheduled operations or damage to property and equipment.
- **Unsafe condition:** This is any condition, under the right set of conditions, that will result in an accident.
- **Unsafe act:** This is an act that is not safe for an employee/person.
- **Safeguard:** This is a barrier guard, device, or procedure developed for protecting humans.
- **Hazard:** This is the source of energy and the physiological and behavioral factors which, when uncontrolled, result in harmful occurrences.
- **Accident:** This is an event that involves damage to a certain system that suddenly disrupts the ongoing or potential system output.
- **Failure:** This is the inability of a system/item to carry out its stated function.

- **Downtime:** This is the time during which the system/item is not in a condition to carry out its specified mission.
- **Mission time:** This is the element of uptime needed for performing a stated mission profile.
- **Redundancy:** This is the existence of more than one means for performing a specified function.
- **Useful life:** This is the length of time a system/item functions within an acceptable failure rate level.
- **Human performance:** This is a measure of actions and failures under stated conditions.
- **Continuous task:** This is a task/job that involves some kind of tracking activity (e.g., monitoring a changing condition).
- **Man function:** This is that function which is allocated to the system's human element.
- **Unsafe behavior:** This is the manner in which an individual carries out actions that are considered unsafe to himself/herself or other people.
- **Human reliability:** This is the probability of accomplishing a task successfully by humans at any required stage in system operations within a stated minimum limit (i.e., if the time requirement is stated).
- **Human error consequence:** This is an undesired consequence of human failure.
- **Risk:** This is a hazardous condition's probable occurrence rate and the degree of harm severity.
- **Maintenance:** This is all actions appropriate for retaining an item/equipment in, or restoring it to, a stated condition.
- **Maintainability:** This is the probability that a failed item/system/equipment will be restored to satisfactory working condition.

1.4 Useful Information on Safety, Reliability, Human Factors, and Human Error in Nuclear Power Plants

This section lists books, journals, technical reports, conference proceedings, data sources, and organizations that are directly or indirectly considered quite useful for obtaining information concerning safety, reliability, human factors, and human error in nuclear power plants.

1.4.1 Books

Some of the books that are directly or indirectly concerned with safety, reliability, human factors, and human error in nuclear power plants are listed below.

- Grigsby, L.E., Editor, *Electric Power Generation, Transmission, and Distribution*, CRC Press, Boca Raton, Florida, 2007.
- Cepin, M., *Assessment of Power System Reliability: Methods and Applications*, Springer, London, 2011.
- Dhillon, B.S., *Power System Reliability, Safety, and Management*, Ann Arbor Science Publishers, Ann Arbor, Michigan, 1983.
- Stephans, R.A., Talso, W.W., Editors, *System Safety Analysis Handbook*, System Safety Society, Irvine, California, 1993.
- Dhillon, B.S., *Engineering Safety: Fundamentals, Techniques, and Applications*, World Scientific Publishing, River Edge, New Jersey, 2003.
- Cox, S.J., *Reliability, Safety, and Risk Management: An Integrated Approach*, Butterworth-Heinemann, New York, 1991.
- Dhillon, B.S., *Design Reliability: Fundamentals and Applications*, CRC Press, Boca Raton, Florida, 1999.
- Salvendy, G., Editor, *Handbook of Human Factors and Ergonomics*, John Wiley & Sons, New York, 2006.
- Oborne, D.J., *Ergonomics at Work: Human Factors in Design and Development*, John Wiley & Sons, New York, 1995.
- Corlett, E.N., Clark, T.S., *The Ergonomics of Workspaces and Machines*, Taylor & Francis, London, 1995.
- Proctor, R.W., Van Zandt, T., *Human Factors in Simple and Complex Systems*, CRC Press, Boca Raton, Florida, 2008.
- Dhillon, B.S., *Human Reliability with Human Factors*, Pergamon Press, New York, 1986.
- Dekker, S., *Ten Questions about Human Error: A New View of Human Factors and System Safety*, Lawrence Erlbaum Associates, Mahwah, New Jersey, 2005.
- Dhillon, B.S., *Human Reliability, Error, and Human Factors in Power Generation*, Springer, London, 2014.
- Sanders, M.S., McCormick, E.J., *Human Factors in Engineering and Design*, McGraw-Hill, New York, 1993.
- Dhillon, B.S., *Human Reliability, Error, and Human Factors in Engineering Maintenance*, CRC Press, Boca Raton, Florida, 2009.

- Strauch, B., *Investigating Human Error: Incidents, Accidents, and Complex Systems*, Ashgate Publishing, Aldershot, UK, 2002.
- Karwowski, M., Marras, W.S., *The Occupational Ergonomics Handbook*, CRC Press, Boca Raton, Florida, 1999.

1.4.2 Journals

Some of the journals that from time to time publish articles that are directly or indirectly concerned with safety, reliability, human factors, and human error in nuclear power plants are listed below.

- *Nuclear Safety*
- *IEEE Transactions on Power Apparatus and Systems*
- *IEEE Transactions on Power Delivery*
- *Reliability Engineering and System Safety*
- *Nuclear Energy and Engineering*
- *Nuclear Engineering and Design*
- *Journal of Korean Nuclear Society*
- *Power Engineering*
- *Nuclear Europe Worldscan*
- *Accident Prevention and Analysis*
- *Journal of Risk and Reliability*
- *IEEE Power and Energy Magazine*
- *International Journal of Reliability, Quality, and Safety Engineering*
- *Electric Power Systems Research*
- *Human Factors*
- *Applied Ergonomics*
- *International Journal of Man-Machine Studies*
- *Human Factors in Aerospace and Safety*
- *IEEE Transactions on Systems, Man, and Cybernetics*
- *Ergonomics*
- *IEEE Transactions on Reliability*
- *Human Factors and Ergonomics in Manufacturing*
- *International Journal of Power and Energy Systems*
- *IEEE Transactions on Industry Applications*
- *Journal of Quality in Maintenance Engineering*
- *Progress in Nuclear Energy*
- *Fusion Engineering and Design*

- *Microelectronics and Reliability*
- *Atomic Energy*
- *IEEE Transactions on Nuclear Science*
- *International Journal of Energy Research*
- *Nuclear Engineer*
- *Transactions of the American Nuclear Society*
- *Safety Science*
- *Journal of Nuclear Science and Technology*
- *Annals of Nuclear Energy*

1.4.3 Conference Proceedings

Some of the conference proceedings that contain articles that are directly or indirectly concerned with safety, reliability, human factors, and human error in nuclear power plants are listed below.

- *Proceedings of the IEEE Conference on Human Factors and Power Plants*
- *Proceedings of the IEEE Conference on Human Factors and Nuclear Safety*
- *Proceedings of the International Topical Meeting on Nuclear Plant Instrumentation, Controls, and Human-Machine Interface Technology*
- *Proceedings of the International Conference on Reliability, Maintainability, and Safety*
- *Proceedings of the Annual Reliability and Maintainability Symposium*
- *Proceedings of the Human Factors and Ergonomics Society Annual Meeting*
- *Proceedings of the International Symposium on Advances in the Operational Safety of Nuclear Power Plants*
- *Proceedings of the International Conference on Advances in Nuclear Power Plants*
- *Proceedings of the International Conference on Nuclear Energy for New Europe*
- *Proceedings of the International Conference on Nuclear Engineering*
- *Proceedings of the IEEE International Conference on Systems, Man, and Cybernetics*
- *Proceedings of the American Nuclear Society International Topical Meeting on Probabilistic Safety Analysis*
- *Proceedings of the International Conference on Reliability, Maintainability, and Safety*
- *Proceedings of the World Congress on Intelligent Control and Automation*
- *Proceedings of the International Conference on Design and Safety of Advanced Nuclear Power Plants*

- *Proceedings of the IEEE International Conference on Human Interfaces in Control Rooms*
- *Proceedings of the European Photo Voltaic Solar Energy Conference and Exhibition*
- *Proceedings of the International Scientific Conference on Electric Power Engineering*

1.4.4 Technical Reports

Some of the important technical reports that are directly or indirectly concerned with safety, reliability, human factors, and human error in nuclear power plants are listed below.

- Trager, T.A., Jr., "Case Study Report on Loss of System Safety Function Events," Report No. AEOD/C504, United States Nuclear Regulatory Commission, Washington, DC, 1985.
- "Nuclear Power Plant Operating Experience," from the IAEA/NEA Incident Reporting System, Report, Organization for Economic Co-operation and Development (OECD), 2 rue Andre-Pascal, 7575 Paris Cedex 16, France, 2000.
- WASH-1400, Reactor Safety Study: An Assessment of Accident Risks in U.S. Commercial Nuclear Power Plants, U.S. Nuclear Regulatory Commission, Washington, DC, 1975.
- "An Analysis of 1990 Significant Events," Report No. INP091-018, Institute of Nuclear Power Operations (INPO), Atlanta, Georgia, 1991.
- "An Analysis of Root Causes in 1983 and 1984 Significant Event Reports," Report No. 85-027, Institute of Nuclear Power Operations, Atlanta, Georgia, July 1985.
- Kolaczkowshi, A., Forester, J., Lois, E., Cooper, S., "Good Practices for Implementing Human Reliability Analysis (HRA)," Report No. NUREG-1792, United States Nuclear Regulatory Commission, Washington, DC, April 2005.
- Maintenance Error Decision Aid (MEDA), Developed by Boeing Commercial Airplane Group, Seattle, Washington, 1994.
- McCornack, R.L., Inspector Accuracy: A Study of the Literature, Report No. SCTM 53-61 (14), Sandia Corporation, Albuquerque, New Mexico, 1961.
- Seminara, J.L., Parsons, S.O., Human Factors Review of Power Plant Maintenance, Report No. EPRI NP-1567, Electric Power Research Institute, Palo Alto, California, 1981.

1.4.5 Data Sources

There are many sources for obtaining safety, reliability, human factors, and human error data. Some of the sources that could be quite useful for obtaining data that are directly or indirectly related to safety, reliability, human factors, and human error in the area of nuclear power generation are listed below.

- Collection and Classification of Human Reliability Data for Use in Probabilistic Safety Assessments, Report No. IAEA-TECDOC-1048, International Atomic Energy Agency (IAEA), Vienna, Austria, 1998.
- Human Error Classification and Data Collection, Report No. IAEA-TEC-DOC-538, International Atomic Energy Agency, Vienna, Austria, 1990.
- Gertman, D.A., Blackman, H.S., *Human Reliability and Safety Analysis Data Handbook*, John Wiley & Sons, New York 1994.
- Government Industry Data Exchange Program (GIDEP), GIDEP Operations Center, U.S. Department of Navy, Corona, California, USA.
- Swain, A.D., Guttman, H.E., Handbook of Human Reliability Analysis with Emphasis on Nuclear Power Plant Applications, Report No. NUREG/CR-1278, U.S. Nuclear Regulatory Commission, Washington, DC, 1983.
- Stewart, C., The Probability of Human Error in Selected Nuclear Maintenance Tasks, Report No. EGG-SSDC-5580, Idaho National Engineering Laboratory, Idaho Falls, Idaho, USA, 1981.
- Dhillon, B.S., *Human Reliability: With Human Factors*, Pergamon Press, New York, 1986. (This book lists over 20 sources for obtaining human reliability-related data).
- Data on Equipment Used in Electric Power Generation, Equipment Reliability Information Center (ERIC), Canadian Electrical Association, Montreal, Quebec, Canada.
- Schmidtke, H., Editor, *Ergonomic Data for Equipment Design*, Plenum Press, New York, 1984.
- Dhillon, B.S., Human Error Data Banks, *Microelectronics and Reliability*, Vol. 30, 1990, pp. 963–971.
- Boff, K.R., Lincoln, J.E., *Engineering Data Compendium: Human Perception and Performance*, Vols. 1–3, Armstrong Aerospace Medical Research Laboratory, Wright-Patterson Air Force Base, Ohio, USA, 1988.

1.4.6 Organizations

There are many organizations that collect safety, reliability, human factors, and human error-related data/information. Some of the organizations that could be quite useful for obtaining data/information directly or indirectly related to safety, reliability, human factors, and human error in nuclear power plants are listed below.

- International System Safety Society, Unionville, Virginia, USA
- American Society of Safety Engineers, 1800 E Oakton Street, Des Plaines, Illinois, USA
- American Nuclear Society, 555 North Kensington Avenue, La Grange Park, Illinois, USA
- IEEE Reliability Society, c/o IEEE Corporate Office, 3 Oak Avenue, 17th Floor, New York, USA
- International Atomic Energy Agency, Wagramer Strasse 5, Vienna, Austria
- Society for Machinery Failure Prevention Technology, 4193 Sudley Road, Haymarket, Virginia, USA
- IEEE Power & Energy Society, 445 Hoes Lane, Piscataway, New Jersey, USA
- Society for Maintenance and Reliability Professionals, 401 N. Michigan Avenue, Chicago, Illinois, USA
- Human Factors and Ergonomics Society, 1124 Montana Avenue, Suite B, Santa Monica, California, USA
- Canadian Nuclear Society, 655 Bay Street, 17th Floor, Toronto, Ontario, Canada

1.5 Scope of the Book

Just like any other area of engineering, the area of nuclear power generation is also subjected to safety, reliability, human factors, and human error-related problems. Nowadays, increasing attention is being given to safety, reliability, human factors, and human error-related problems in the area of nuclear power generation due to various factors, including cost and serious consequences such as the Chernobyl nuclear accident in Ukraine and the Three Mile Island nuclear accident in the United States.

Over the years, a large number of publications that are directly or indirectly related to safety, reliability, human factors, and human error in nuclear power plants have appeared. Almost all of these publications are in the form of journal or conference proceedings articles or technical reports. At present, to the best of the author's knowledge, there is no specific book that covers the topic of this book within its framework. This book attempts to provide up-to-date coverage not only of the ongoing effort in safety, reliability, human factors, and human error in nuclear power plants but also of useful developments in the general areas of safety, reliability, human factors, and human error.

Finally, the main objective of this book is to provide professionals concerned with safety, reliability, human factors, and human error in nuclear power plants information that could be useful to eradicate or reduce the occurrence of safety, reliability, human factors, and human error-related problems in this area. The book will be useful to many individuals including engineering professionals working in the area of nuclear power generation; researchers and instructors involved with nuclear power plant systems; safety, reliability, human factors, and human error professionals and administrators involved with nuclear power plants; and graduate students in the area of nuclear power generation and reliability and safety engineering.

1.6 Problems

1. Write an essay on safety, reliability, human factors, and human error in nuclear power plants.
2. List at least seven facts, figures, and examples that are directly or indirectly concerned with safety, reliability, human factors, and human error in nuclear power plants.
3. Define the following four terms:
   a. Safety
   b. Reliability
   c. Human factors
   d. Human error
4. List at least five sources for obtaining human error and reliability in nuclear power generation-related data.
5. Compare the terms "unsafe act" and "unsafe behavior."
6. Define the following terms:
   a. Hazard
   b. Accident
   c. Human reliability
7. List at least seven important books for obtaining information that is directly or indirectly related to safety, reliability, human factors, and human error in nuclear power plants.

8. List four of the most important organizations for obtaining information that is directly or indirectly related to safety, reliability, human factors, and human error in nuclear power plants.

9. Define the following three terms:
   a. Man function
   b. Risk
   c. Continuous task

10. List seven of the most important journals for obtaining information related to safety, reliability, human factors, and human error in nuclear power plants.

References

1. Kessler, G., *Sustainable and Safe Nuclear Fission Energy*, Power Systems, Springer-Verlag, Berlin, 2012.
2. Facts and Figures, *Nuclear Industry Association (NIA)*, Carleton House, London, 2013.
3. Varma, V., Maintenance training reduces human errors, *Power Eng.*, Vol. 100, 1996, pp. 44–47.
4. Trager, T.A., Jr., Case Study Report on Loss of Safety System Function Events, Report No. AEOD/C 504, United States Nuclear Regulatory Commission (NRC), Washington, DC, 1985.
5. An Analysis of Root Causes in 1983 and 1984 Significant Event Reports, Report No. 85-027, Institute of Nuclear Power Operations (INPO), Atlanta, Georgia, July 1985.
6. Williams, J.C., A data-based method for assessing and reducing human error to improve operational performance, *Proceedings of the IEEE Conference on Human Factors and Power Plants*, 1988, pp. 436–450.
7. Mishima, S., Human factors research program: Long term plan in cooperation with government and private research centers, *Proceedings of the IEEE Conference on Human Factors and Power Plants*, 1988, pp. 50–54.
8. Muschara, T.M., Eliminating plant events by reducing the number of shots on goal for IEEE conference record on power engineering, *Proceedings of the IEEE Sixth Annual Human Factors Meeting*, 1997, pp. 12.1–12.6.
9. Pyy, P., Laakso, K., Reiman, L., A study of human errors related to NPP maintenance activities, *Proceedings of the IEEE 6th Annual Human Factors Meeting*, 1997, pp. 12.23–12.28.
10. Pyy, P., A analysis of maintenance failures at a nuclear power plant, *Reliability Engineering and System Safety*, Vol. 72, 2001, pp. 293–302.
11. Daniels, R.W., The formula for improved plant maintainability must include human factors, *Proceedings of the IEEE Conference on Human Factors and Nuclear Safety*, 1985, pp. 242–244.
12. Scott, R.L., Recent occurrences of nuclear reactors and their causes, *Nuclear Safety*, Vol. 16, 1975, pp. 496–497.

13. Husseiny, A.A., Sabry, I.A., Analysis of human factor in operation of nuclear power plants, *Atomkernenergie Kerntechnik*, Vol. 36, 1980, pp. 115–121.
14. Heo, G., Park, J., A framework for evaluating the effects of maintenance-related human errors in nuclear power plants, *Reliability Engineering and System Safety*, Vol. 95, 2010, pp. 797–805.
15. Lee, J.W., Park, G.O., Park, J.C., Sim, B.S., Analysis of error trips cases in Korean NPPs, *Journal of Korean Nuclear Society*, Vol. 28, 1996, pp. 563–575.
16. Kim, J., Park, J., Jung, W., Kim, J.T., Characteristics of test and maintenance human errors leading to unplanned reactor trips in nuclear power plants, *Nuclear Engineering and Design*, Vol. 239, 2009, pp. 2530–2536.
17. *Maintenance Error a Factor in Blackouts, Miami Herald*, Miami, Florida, December 29, 1989, p. 4.
18. Reason, J., Human factors in nuclear power generation: A systems perspective, *Nuclear Europe Worldscan*, Vol. 17, No. 5–6, 1997, pp. 35–36.
19. Kawano, R., Steps toward the realization of human-centered systems: An overview of the human factors activities at TEPCO, *Proceedings of the IEEE Sixth Annual Human Factors Meeting*, 1997, pp. 13.27–13.32.
20. Heo, G., Park, J., Framework of quantifying human error effects in testing and maintenance, *Proceedings of the Sixth American Nuclear Society International Topical Meeting on Nuclear Plant Instrumentation, Control, and Human-Machine Interface Technologies*, 2009, pp. 2083–2092.
21. Dhillon, B.S., *Human Reliability, Error, and Human Factors in Power Generation*, Springer, London, 2014.
22. Omdahl, T.P., ed., *Reliability, Availability, and Maintainability (RAM) Dictionary*, ASQC Quality Press, Milwaukee, 1988.
23. McKenna, T., Oliverson, R., *Glossary of Reliability and Maintenance Terms*, Gulf Publishing Company, Houston, Texas, 1997.
24. MIL-STD-721C, *Definitions of Terms for Reliability and Maintainability*, Department of Defense, Washington, DC.
25. Naresky, J.J., Reliability definitions, *IEEE Transactions on Reliability*, Vol. 19, 1970, pp. 198–200.
26. Von Alven, W.H., ed, *Reliability Engineering*, Prentice Hall, Englewood Cliffs, New Jersey, 1964.
27. Meister, D., Human factors in reliability, in *Reliability Handbook*, edited by W.G. Ireson, McGraw-Hill, New York, 1966, pp. 12.2–12.37.
28. MIL-STD-721B, *Definitions of Effectiveness Terms for Reliability, Maintainability, Human Factors, and Safety*, Department of Defense, Washington, DC, August 1966. Available from the Naval Publications and Forms Center, 5801 Tabor Avenue, Philadelphia, Pennsylvania.
29. MIL-STD-1908, *Definitions of Human Factors Terms*, Department of Defense, Washington, DC.
30. *Dictionary of Terms Used in the Safety Profession*, 3rd Edition, American Society of Safety Engineers (ASSE), Des Plaines, Illinois, 1988.

2

Basic Mathematical Concepts

2.1 Introduction

As in the development of other areas of science and technology, mathematics has also played an important role in the development of the safety, reliability, and human error fields in the area of nuclear power plants. The history of mathematics may be traced back to the development of our currently used number symbols, sometimes in the published literature referred to as the "Hindu-Arabic numeral system" [1]. The first evidence of the use of these symbols is found on stone columns erected by the Scythian emperor of India named Asoka in around 250 BCE [1].

The earliest reference to the concept of probability may be traced back to a gambler's manual written by Girolamo Cardano (1501–1576) [2]. However, Pierre Fermat (1601–1665) and Blaise Pascal (1623–1662) were the first two persons who solved the problem of dividing the winnings in a game of chance independently and correctly [1,2]. Boolean algebra, which plays an important role in modern probability theory, is named after the English mathematician George Boole (1815–1864), who published a pamphlet entitled "The Mathematical Analysis of Logic: Being an Essay Towards a Calculus of Deductive Reasoning" in 1847 [1–3].

Laplace transforms, often used in the area of reliability for finding solutions to first-order linear differential equations, were developed by Pierre-Simon Laplace (1749–1827), a French mathematician. A more detailed history of mathematics and probability is available in References 1 and 2. This chapter presents basic mathematical concepts considered useful for understanding the subsequent chapters of this book.

2.2 Arithmetic Mean and Mean Deviation

A given set of data that are directly or indirectly related to safety, reliability, human factors, and human error concerning engineering systems used in nuclear power plants is useful only if it is analyzed properly. More specifi-cally, there are certain characteristics of the data that are useful for describ-ing the nature of a given data set, thus making more effective associated decision.

Thus, this section presents two statistical measures considered quite use-ful to study safety, reliability, human factors, and human error-related data concerning systems used in nuclear power plants [4–7].

2.2.1 Arithmetic Mean

Often, this is simply referred to as mean and is expressed by

$$m = \frac{\sum_{i=1}^{k} y_i}{k} \tag{2.1}$$

where
  k is the number of data values.
  y_i is the data value i, for $i = 1, 2, 3, ..., k$.
  m is the mean value (i.e., arithmetic mean).

EXAMPLE 2.1

Assume that the quality control department of a nuclear power plant equipment manufacturing company inspected seven identical sys-tems and discovered 2,4,1,3,1,2, and 1 defects in each respective sys-tem. Calculate the average number of defects (i.e., arithmetic mean) per system.

By substituting the specified data values into Equation 2.1, we obtain:

$$m = \frac{2+4+1+3+1+2+1}{7} = 2$$

Thus, the average number of defects per system is 2. More specifically, the arithmetic mean of the given data set is 2.

2.2.2 Mean Deviation

This is a measure of dispersion whose value indicates the degree to which a given set of data tends to spread about a mean value. Mean deviation is defined by

$$MD = \frac{\sum_{i=1}^{n} |y_i - m|}{n} \tag{2.2}$$

where
   n is the number of data values.
   y_i is the data value i, for $i = 1, 2, 3, \ldots, n$.
   MD is the mean deviation.
   m is the mean value of the given data set.
   $|y_i - m|$ is the absolute value of the deviation of y_i from m.

EXAMPLE 2.2

Calculate the mean deviation of the data set given in Example 2.1.
   In Example 2.1, the calculated mean value (i.e., arithmetic mean) of the given data set is 2 defects per system. Thus, by using this calculated value and the given data values in Equation 2.2, we get:

$$MD = \frac{|2-2|+|4-2|+|1-2|+|3-2|+|1-2|+|2-2|+|1-2|}{7}$$

$$= \frac{[0+2+1+1+1+0+1]}{7}$$

$$= 0.857$$

Thus, the mean deviation of the data set given in Example 2.1 is 0.857.

2.3 Boolean Algebra Laws

Boolean algebra is used to a degree in various studies concerning safety, reliability, human factors, and human error in nuclear power plants and is named after its founder, George Boole (1813–1864) [3]. Some of its laws that are considered useful to understanding subsequent chapters of this book are presented below [3–9].

$$A \cdot B = B \cdot A \tag{2.3}$$

where
   Dot(·) denotes the intersection of sets or events.
   A is an arbitrary set or event.
   B is an arbitrary set or event.

It should be noted that many times Equation 2.3 is written without the dot (e.g., AB), but it still conveys the same meaning.

$$A + B = B + A \tag{2.4}$$

where
   + denotes the union of sets or events.

$$A + A = A \tag{2.5}$$

$$AA = A \tag{2.6}$$

$$A + AB = A \tag{2.7}$$

$$B(B + A) = B \tag{2.8}$$

$$A(B + C) = AB + AC \tag{2.9}$$

where
C is an arbitrary set or event.

$$(A + B)(A + C) = A + BC \tag{2.10}$$

$$(A + B) + C = A + (B + C) \tag{2.11}$$

$$(AB)C = A(BC) \tag{2.12}$$

It should be noted that in the published literature, Equations 2.3 and 2.4 are referred to as commutative law, Equations 2.11 and 2.12 as associative law, Equations 2.9 and 2.10 as distributive law, Equations 2.7 and 2.8 as absorption law, and Equations 2.5 and 2.6 as idempotent law [10].

2.4 Probability Definition and Properties

The probability is defined as follows [11]:

$$P(Y) = \lim_{n \to \infty} \left[\frac{N}{n} \right] \tag{2.13}$$

where
$P(Y)$ is the probability of occurrence of event Y.
N is the number of times event Y occurs in the n repeated experiments.

Some of the basic properties of probability are as follows [8,11]:

- The probability of occurrence of an event, say event Y, is

$$0 \le P(Y) \le 1 \tag{2.14}$$

- The probability of occurrence and nonoccurrence of an event, say event Y, is always

$$P(Y) + P(\overline{Y}) = 1 \tag{2.15}$$

where
  $P(Y)$ is the probability of occurrence of event Y.
  $P(\overline{Y})$ is the probability of nonoccurrence of event Y.

- The probability of the union of k-independent events is

$$P(Y_1 + Y_2 + \cdots + Y_k) = 1 - \prod_{i=1}^{k}(1 - P(Y_i)) \tag{2.16}$$

where
  $P(Y_i)$ is the probability of occurrence of event Y_i, for $i = 1, 2, 3, \ldots, k$.

- The probability of the union of k mutually exclusive events is

$$P(Y_1 + Y_2 + \cdots + Y_k) = \sum_{i=1}^{k} P(Y_i) \tag{2.17}$$

- The probability of an interaction of k-independent events is

$$P(Y_1 \ Y_2 \ Y_3 \ldots Y_k) = P(Y_1)P(Y_2)P(Y_3)\ldots P(Y_k) \tag{2.18}$$

EXAMPLE 2.3

Assume that a system used in a nuclear power plant is composed of two very critical subsystems, say subsystem Y_1 and subsystem Y_2. The failure of either subsystem can directly or indirectly result in system failure. The failure probability of subsystems Y_1 and Y_2 is 0.05 and 0.08, respectively. Calculate the probability of failure of the system if both of these subsystems fail independently.

By inserting the given data values into Equation 2.16, we obtain

$$
\begin{aligned}
P(Y_1 + Y_2) &= 1 - \prod_{i=1}^{2}(1 - P(Y_i)) \\
&= P(Y_1) + P(Y_2) - P(Y_1)P(Y_2) \\
&= 0.05 + 0.08 - (0.05)(0.08) \\
&= 0.126
\end{aligned}
$$

Thus, the probability of failure of the system is 0.126.

2.5 Useful Mathematical Definitions

This section presents five mathematical definitions considered useful to performing various types of safety, reliability, human factors, and human error studies that are directly or indirectly concerned with systems used in nuclear power plants.

2.5.1 Definition I: Cumulative Distribution Function

For a continuous random variable, the cumulative distribution function is defined by [6,11]

$$F(t) = \int_{-\infty}^{t} f(x)\,dx \tag{2.19}$$

where
  $F(t)$ is the cumulative distribution function.
  x is a continuous random variable.
  $f(x)$ is the probability density function.

  For $t = \infty$, Equation 2.19 becomes

$$F(\infty) = \int_{-\infty}^{\infty} f(x)\,dx \tag{2.20}$$
$$= 1$$

It simply means that the total area under the probability density curve is equal to unity. Generally, in reliability, safety, and human error studies of systems, Equation 2.19 is simply written as

$$F(t) = \int_{0}^{t} f(x)\,dx \tag{2.21}$$

EXAMPLE 2.4

Assume that the probability (i.e., failure) density function of a system used in a nuclear power plant is

$$f(t) = \lambda_s e^{-\lambda_s t} \quad \text{for } t \leq 0, \ \lambda_s > 0 \tag{2.22}$$

where
  t is a continuous random variable (i.e., time).

$f(t)$ is the probability density function (generally, in the area of reliability engineering, it is called the failure density function).
λ_s is the system failure rate.

Obtain an expression for the system cumulative distribution function. By inserting Equation 2.22 into Equation 2.21, we get

$$F(t) = \int_0^t \lambda_s e^{-\lambda_s t} = 1 - e^{-\lambda_s t} \qquad (2.23)$$

Thus, Equation 2.23 is the expression for the system cumulative distribution function.

2.5.2 Definition II: Probability Density Function

For a continuous random variable, the probability density function is expressed by [11]

$$f(t) = \frac{dF(t)}{dt} \qquad (2.24)$$

where
$f(t)$ is the probability density function.
$F(t)$ is the cumulative distribution function.

EXAMPLE 2.5

Prove by using Equation 2.23 that Equation 2.22 is the probability density function.
By inserting Equation 2.23 into Equation 2.24, we get

$$f(t) = \frac{d(1 - e^{-\lambda_s t})}{dt} = \lambda_s e^{-\lambda_s t} \qquad (2.25)$$

Equations 2.25 and 2.22 are identical.

2.5.3 Definition III: Expected Value

The expected value of a continuous random variable is expressed by [11]

$$E(t) = \int_{-\infty}^{\infty} t f(t) dt \qquad (2.26)$$

where
$E(t)$ is the expected value (i.e., mean value) of the continuous random variable t.

EXAMPLE 2.6

Find the expected value (i.e., mean value) of the probability (failure) density function expressed by Equation 2.22.

By substituting Equation 2.22 into Equation 2.26, we obtain

$$E(t) = \int_0^\infty t \lambda_s e^{-\lambda_s t} dt$$

$$= [-t e^{-\lambda_s t}]_0^\infty - \left[-\frac{e^{-\lambda_s t}}{\lambda_s} \right]_0^\infty \qquad (2.27)$$

$$= \frac{1}{\lambda_s}$$

Thus, the expected value (i.e., mean value) of the probability (failure) density function expressed by Equation 2.22 is given by Equation 2.27.

2.5.4 Definition IV: Laplace Transform

This is named after a French mathematician, Pierre-Simon Laplace (1749–1827), and is defined by [1,12,13]

$$f(s) = \int_0^\infty f(t) e^{-st} dt \qquad (2.28)$$

where

t is a variable.
s is the Laplace transform variable.
$f(s)$ is the Laplace transform of function $f(t)$.

An example of obtaining Laplace transform by using Equation 2.28 is presented in the following, and Laplace transforms of some commonly occurring functions in nuclear power plant systems safety, reliability, and human error-related analysis studies are presented in Table 2.1 [12–14].

EXAMPLE 2.7

Obtain the Laplace transform of the following function:

$$f(t) = e^{-\mu t} \qquad (2.29)$$

where

μ is a constant.

TABLE 2.1

Laplace Transforms of Some Functions

No.	$f(t)$	$f(s)$
1	t^k, for $k = 0, 1, 2, 3, \ldots$	$\dfrac{k!}{s^{k+1}}$
2	t	$\dfrac{1}{s^2}$
3	C (a constant)	$\dfrac{C}{s}$
4	$e^{-\mu t}$	$\dfrac{1}{s+\mu}$
5	$\gamma_1 f_1(t) + \gamma_2 f_2(t)$	$\gamma_1 f_1(s) + \gamma_2 f_2(s)$
6	$\dfrac{df(t)}{dt}$	$s f(s) - f(0)$
7	$te^{-\mu t}$	$\dfrac{1}{(s+\mu)^2}$
8	$t f(t)$	$-\dfrac{df(s)}{ds}$

By inserting Equation 2.29 into Equation 2.28, we obtain

$$f(s) = \int_0^\infty e^{-\mu t} e^{-st} dt$$

$$= \frac{e^{-(s+\mu)t}}{(s+\mu)} \Big|_0^\infty \tag{2.30}$$

$$= \frac{1}{s+\mu}$$

Thus, Equation 2.30 is the Laplace transform of Equation 2.29.

2.5.5 Definition V: Final Value Theorem Laplace Transform

If the following limits exist, then the final value theorem may be stated as

$$\lim_{t \to \infty} f(t) = \lim_{s \to 0} sf(s) \tag{2.31}$$

EXAMPLE 2.8

Prove, by using the following equation, that the left-hand side of Equation 2.31 is equal to its right-hand side:

$$f(t) = \frac{\gamma_1}{(\gamma_1 + \gamma_2)} + \frac{\gamma_2}{(\gamma_1 + \gamma_2)} e^{-(\gamma_1 + \gamma_2)t} \tag{2.32}$$

where
γ_1 and γ_2 are constants.

By inserting Equation 2.32 into the left-hand side of Equation 2.31, we obtain

$$\lim_{t\to\infty}\left[\frac{\gamma_1}{(\gamma_1+\gamma_2)}+\frac{\gamma_2}{(\gamma_1+\gamma_2)}e^{-(\gamma_1+\gamma_2)t}\right]=\frac{\gamma_1}{(\gamma_1+\gamma_2)} \qquad (2.33)$$

By using Table 2.1, we get the following Laplace transforms of Equation 2.32:

$$f(s)=\frac{\gamma_1}{s(\gamma_1+\gamma_2)}+\frac{\gamma_2}{(\gamma_1+\gamma_2)}\cdot\frac{1}{(s+\gamma_1+\gamma_2)} \qquad (2.34)$$

By substituting Equation 2.34 into the right-hand side of Equation 2.31, we obtain

$$\lim_{s\to0}\left[\frac{s\gamma_1}{s(\gamma_1+\gamma_2)}+\frac{s\gamma_2}{(\gamma_1+\gamma_2)(s+\gamma_1+\gamma_2)}\right]=\frac{\gamma_1}{\gamma_1+\gamma_2} \qquad (2.35)$$

As the right-hand sides of Equations 2.33 and 2.35 are the same, it proves that the left-hand side of Equation 2.31 is equal to its right-hand side.

2.6 Probability Distributions

Although there are a large number of probability/statistical distributions, this section presents just five such distributions considered useful for performing various types of nuclear power plant system studies that are directly or indirectly related to safety, reliability, human factors, and human error [15–18].

2.6.1 Binomial Distribution

This is a discrete random variable probability distribution and is used in circumstances in which one is concerned with the probabilities of outcome such as the number of occurrences (e.g., failures) in a sequence of stated number of trials. More specifically, each trial has two possible outcomes (e.g., success or failure), but the probability of each trial remains unchanged/constant. It should be noted that this distribution is also known as the Bernoulli distribution after its founder, Jakob Bernoulli (1654–1705) [1]. The distribution probability density function is expressed by

$$f(y) = \frac{n!}{y!(n-y)!} p^y q^{n-y} \quad \text{for } y = 0, 1, 2, 3, \ldots, n. \tag{2.36}$$

where
  y is the number of nonoccurrences (e.g., failures) in n trials.
  p is the single trial probability of occurrence (e.g., success).
  q is the single trial probability of nonoccurrence (e.g., failure).

The cumulative distribution function is defined by

$$F(y) = \sum_{i=0}^{y} \frac{n!}{i!(n-i)!} p^i q^{n-i} \tag{2.37}$$

where
  $F(y)$ is the probability of y or less nonoccurrences (e.g., failures) in n trials.

2.6.2 Exponential Distribution

This is one of the simplest continuous random variable probability distributions frequently used in the industrial sector, particularly to perform various types of reliability-related studies. Its probability density function is expressed by [19].

$$f(t) = \mu e^{-\mu t} \quad \text{for } \mu > 0, \ t \geq 0 \tag{2.38}$$

where
  μ is the distribution parameter.
  t is the time (i.e., continuous random variable).
  $f(t)$ is the probability density function.

By substituting Equation 2.38 into Equation 2.21, we get the following expression for the cumulative distribution function:

$$F(t) = 1 - e^{-\mu t} \tag{2.39}$$

By using Equations 2.26 and 2.38, we obtain the following expression for the distribution expected value (i.e., mean value):

$$E(t) = \frac{1}{\mu} \tag{2.40}$$

EXAMPLE 2.9

Assume that the mean time to failure of a system used in a nuclear power plant is 3000 hours. Calculate the probability of failure of the system during a 600-hour mission with the aid of Equations 2.39 and 2.40.

By substituting the specified data value into Equation 2.40, we get

$$\mu = \frac{1}{3000} = 0.00033 \text{ failures per hour}$$

By inserting the calculated and the specified data values into Equation 2.39, we get

$$F(600) = 1 - e^{-(0.00033)(600)}$$
$$= 0.1813$$

Thus, the probability of failure of the system during the 600-hour mission is 0.1813.

2.6.3 Rayleigh Distribution

This continuous random variable probability distribution is named after its founder, John Rayleigh (1842–1919) [1]. The distribution probability density function is defined by

$$f(t) = \left(\frac{1}{\alpha^2}\right) t e^{-(t/\alpha)^2} \quad \text{for } \alpha > 0, \ t \geq 0 \tag{2.41}$$

where

t is the time (i.e., a continuous random variable).
α is the distribution parameter.

By substituting Equation 2.41 into Equation 2.21, we get the following equation for the cumulative distribution function:

$$F(t) = 1 - e^{-(t/\alpha)^2} \tag{2.42}$$

By substituting Equation 2.41 into Equation 2.26, we get the following equation for the distribution expected value (i.e., mean value):

$$E(t) = \alpha \Gamma\left(\frac{3}{2}\right) \tag{2.43}$$

where

$\Gamma(.)$ is the gamma function and is defined by

$$\Gamma(k) = \int_0^\infty t^{k-1}e^{-t}dt \quad \text{for } k > 0 \tag{2.44}$$

2.6.4 Weibull Distribution

This continuous random variable probability distribution was developed in the early 1950s by Waloddi Weibull, a Swedish professor of mechanical engineering [20]. The distribution's probability density function is defined by

$$f(t) = \frac{at^{a-1}}{\alpha^a}e^{-(t/\alpha)^a} \quad \text{for } a > 0, \alpha > 0, t \geq 0 \tag{2.45}$$

where

t is the time (i.e., a continuous random variable).

α and a are the distribution scale and shape parameters, respectively.

By inserting Equation 2.45 into Equation 2.21, we get the following equation for the cumulative distribution function:

$$F(t) = 1 - e^{-(t/\alpha)^a} \tag{2.46}$$

By substituting Equation 2.45 into Equation 2.26, we get the following equation for the distribution expected value (i.e., mean value):

$$E(t) = \alpha\Gamma\left(1 + \frac{1}{a}\right) \tag{2.47}$$

It should be noted that for $a = 1$ and $a = 2$, the exponential and Rayleigh distributions are the special cases of this distribution, respectively.

2.6.5 Bathtub Hazard Rate Curve Distribution

This continuous random variable probability distribution was developed in 1981 [21]. In the published literature by other authors around the world, it is generally referred to as the Dhillon distribution/model/law [22–44]. The distribution can represent bathtub-shaped, decreasing, and increasing hazard rates of systems used in nuclear power plants.

The probability density function of the distribution is defined by [21]

$$f(t) = k\alpha(\alpha t)^{k-1}e^{-\left\{e^{(\alpha t)^k}-(\alpha t)^k-1\right\}} \quad \text{for } \alpha > 0, k > 0, t \geq 0 \qquad (2.48)$$

where
  t is the time (i.e., a continuous random variable).
  k and α are the distribution shape and scale parameters, respectively.

By substituting Equation 2.48 into Equation 2.21, we get the following equation for the cumulative distribution function:

$$F(t) = 1 - e^{-\left\{e^{(\alpha t)^k}-1\right\}} \qquad (2.49)$$

It should be noted that for $k = 0.5$, the probability distribution gives the bathtub-shaped hazard rate curve, and for $k = 1$, it gives the extreme value distribution. More clearly, at $k = 1$, the extreme value probability distribution is the special case of this probability distribution.

2.7 Solving First-Order Differential Equations Using Laplace Transforms

Generally, Laplace transforms are used for finding solutions to first-order linear differential equations in safety, reliability, and human error analysis-related studies of systems used in nuclear power plants. The following example demonstrates the finding of solutions to a set of linear first-order differential equations, describing a system used in a nuclear power plant with respect to reliability, safety, and human error by using Laplace transforms.

EXAMPLE 2.10

Assume that a system used in a nuclear power plant can be in any of these three states: operating normally, failed safely, or failed unsafely due to human error. The following three first-order linear differential equations describe the system under consideration:

$$\frac{dP_0}{dt} + (\lambda_s + \lambda_{uh})P_0(t) = 0 \qquad (2.50)$$

$$\frac{dP_1(t)}{dt} - \lambda_s P_0(t) = 0 \qquad (2.51)$$

$$\frac{dP_2(t)}{dt} - \lambda_{uh}P_0(t) = 0 \tag{2.52}$$

where

λ_s is the system constant safe failure rate.

λ_{uh} is the system constant unsafe failure rate due to human error.

$P_j(t)$ is the probability that the system is in state j at time t, for $j = 0$ (operating normally), $j = 1$ (failed safely), and $j = 2$ (failed unsafely due to human error).

At time $t = 0$, $P_0(0) = 1$, $P_1(0) = 0$, and $P_2(0) = 0$.

Solve differential Equations 2.50 through 2.52 with the aid of Laplace transforms.

By using Table 2.1, differential Equations 2.50 through 2.52, and the given initial conditions, we obtain

$$sP_0(s) - 1 + (\lambda_s + \lambda_{uh})P_0(s) = 0 \tag{2.53}$$

$$sP_1(s) - \lambda_s P_0(s) = 0 \tag{2.54}$$

$$sP_2(s) - \lambda_{uh}P_0(s) = 0 \tag{2.55}$$

By solving Equations 2.53 through 2.55, we get

$$P_0(s) = \frac{1}{(s + \lambda_s + \lambda_{uh})} \tag{2.56}$$

$$P_1(s) = \frac{\lambda_s}{s(s + \lambda_s + \lambda_{uh})} \tag{2.57}$$

$$P_2(s) = \frac{\lambda_{uh}}{s(s + \lambda_s + \lambda_{uh})} \tag{2.58}$$

By taking the inverse Laplace transforms of Equations 2.56 through 2.58, we get

$$P_0(t) = e^{-(\lambda_s + \lambda_{uh})t} \tag{2.59}$$

$$P_1(t) = \frac{\lambda_s}{(\lambda_s + \lambda_{uh})} \left[1 - e^{-(\lambda_s + \lambda_{uh})t}\right] \tag{2.60}$$

$$P_2(t) = \frac{\lambda_{uh}}{(\lambda_s + \lambda_{uh})} \left[1 - e^{-(\lambda_s + \lambda_{uh})t}\right] \tag{2.61}$$

Thus, Equations 2.59 through 2.61 are the solutions to differential Equations 2.50 through 2.52.

2.8 Problems

1. Assume that the quality control department of a company that manufactures systems for use in nuclear power plants inspected five identical systems and found 2, 4, 3, 6, and 1 defects in each respective system. Calculate the average number of defects (i.e., arithmetic mean) per system.

2. Calculate the mean deviation of the data set given in question 1.

3. What is idempotent law?

4. Define the following two items:
   a. Probability
   b. Cumulative distribution function

5. What are the basic probability properties?

6. Define the following two items:
   a. Probability density function
   b. Laplace transform

7. Write the probability density function for the Weibull distribution. What are the special case distributions of the Weibull distribution?

8. Prove Equations 2.59 through 2.61 with the aid of Equations 2.56 through 2.58. What is the sum of Equations 2.59 through 2.61?

9. Write the probability density function for the bathtub hazard rate curve distribution. What is the special case distribution of this distribution?

10. Prove Equation 2.48 by using Equation 2.49.

References

1. Eves, H., *An Introduction to the History of Mathematics*, Holt, Reinhart, and Winston, New York, 1976.
2. Owen, D.B., ed., *On the History of Statistics and Probability*, Marcel Dekker, New York, 1976.
3. Lipschutz, S., *Set Theory and Related Topics*, McGraw-Hill, New York, 1964.
4. Speigel, M.R., *Probability and Statistics*, McGraw-Hill, New York, 1975.
5. Speigel, M.R., *Statistics*, McGraw-Hill, New York, 1961.
6. Dhillon, B.S., *Robot System Reliability and Safety: A Modern Approach*, CRC Press, Boca Raton, Florida, 2015.
7. Dhillon, B.S., *Reliability, Quality, and Safety for Engineers*, CRC Press, Boca Raton, Florida, 2004.

8. Lipschutz, S., *Probability*, McGraw-Hill, New York, 1965.
9. Fault Tree Handbook, Report No. NUREG-0492, U.S. Nuclear Regulatory Commission, Washington, DC, 1981.
10. Dhillon, B.S., *Computer System Reliability: Safety and Usability*, CRC Press, Boca Raton, Florida, 2013.
11. Mann, N.R., Schafer, R.E., Singpurwalla, N.P., *Methods for Statistical Analysis of Reliability and Life Data*, John Wiley & Sons, New York, 1974.
12. Spiegel, M.R., *Laplace Transforms*, McGraw-Hill, New York, 1965.
13. Oberhettinger, F., Badii, L., *Tables of Laplace Transforms*, Springer-Verlag, Inc., 1973.
14. Nixon, F.E., *Handbook of Laplace Transformation: Fundamentals, Applications, Tables, and Examples*, Prentice Hall, Inc., Englewood Cliffs, New Jersey, 1960.
15. Shooman, M.L., *Probabilistic Reliability: An Engineering Approach*, McGraw-Hill, New York, 1968.
16. Patel, J.K., Kapadia, C.H., Owen, D.H., *Handbook of Statistical Distributions*, Marcel Dekker, New York, 1976.
17. Dhillon, B.S., *Reliability Engineering in Systems Design and Operation*, Van Nostrand Reinhold, New York, 1983.
18. Dhillon, B.S., *Design Reliability: Fundamentals and Applications*, CRC Press, Boca Raton, Florida, 1999.
19. Davis, D.J.: Analysis of some failure data, *Journal of the American Statistical Association*. Vol. 57, 1952, pp. 113–150.
20. Weibull, W.: A statistical distribution function of wide applicability, *Journal of Applied Mechanics*, Vol. 18, 1951, pp. 293–297.
21. Dhillon, B.S., Life distributions, *IEEE Transactions on Reliability*, Vol. 30, 1981, pp. 457–460.
22. Baker, R.D., Nonparametric estimation of the renewal function, *Computers Operations Research*, Vol. 20, No. 2, 1993, pp. 167–178.
23. Cabana, A., Cabana, E.M., Goodness-of-fit to the exponential distribution, focused on Weibull alternatives, *Communications in Statistics-Simulation and Computation*, Vol. 34, 2005, pp. 711–723.
24. Grane, A., Fortiana, J., A directional test of exponentiality based on maximum correlations, *Metrika*, Vol. 73, 2011, pp. 255–274.
25. Henze, N., Meintnis, S.G., Recent and classical tests for exponentiality: A partial review with comparisons, *Metrika*, Vol. 61, 2005, pp. 29–45.
26. Jammalamadaka, S.R., Taufer, E., Testing exponentiality by comparing the empirical distribution function of the normalized spacings with that of the original data, *Journal of Nonparametric Statistics*, Vol. 15, No. 6, 2003, pp. 719–729.
27. Hollander, M., Laird, G., Song, K.S., Nonparametric interference for the proportionality function in the random censorship model, *Nonparametric Statistics*, Vol. 15, No. 2, 2003, pp. 151–169.
28. Jammalamadaka, S.R., Taufer, E., Use of mean residual life in testing departures from exponentiality, *Journal of Nonparametric Statistics*, Vol. 18, No. 3, 2006, pp. 277–292.
29. Kunitz, H., Pamme, H., The Mixed gamma ageing model in life data analysis, *Statistical Papers*, Vol. 34, 1993, pp. 303–318.
30. Kunitz, H., A new class of bathtub-shaped hazard rates and its application in comparison of two test-statistics, *IEEE Transactions on Reliability*, Vol. 38, No. 3, 1989, pp. 351–354.

31. Meintanis, S.G., A class of tests for exponentiality based on a continuum of moment conditions, *Kybernetika*, Vol. 45, No. 6, 2009, pp. 946–959.
32. Morris, K., Szynal, D., Goodness-of-fit tests based on characterizations involving moments of order statistics, *International Journal of Pure and Applied Mathematics*, Vol. 38, No. 1, 2007, pp. 83–121.
33. Na, M.H., Spline hazard rate estimation using censored data, *Journal of KSIAM*, Vol. 3, No. 2, 1999, pp. 99–106.
34. Morris, K., Szynal, D., Some U-statistics in goodness-of-fit tests derived from characterizations via record values, *International Journal of Pure and Applied Mathematics*, Vol. 46, No. 4, 2008, pp. 339–414.
35. Nam, K.H., Park, D.H., Failure rate for Dhillon model, *Proceedings of the Spring Conference of the Korean Statistical Society*, 1997, pp. 114–118.
36. Nimoto, N., Zitikis, R., The Atkinson Index, the moran statistic, and testing exponentiality, *Journal of the Japan Statistical Society*, Vol. 38, No. 2, 2008, pp. 187–205.
37. Nam, K.H., Chang, S.J., Approximation of the renewal function for Hjorth model and Dhillon model, *Journal of the Korean Society for Quality Management*, Vol. 34, No. 1, 2006, pp. 34–39.
38. Noughabi, H.A., Arghami, N.R., Testing exponentiality based on characterizations of the exponential distribution, *Journal of Statistical Computation and Simulation*, Vol. 1, First, 2011, pp. 1–11.
39. Szynal, D., Goodness-of-fit tests derived from characterizations of continuous distributions, *Stability in Probability, Banach Center Publications, Vol. 90*, Institute of Mathematics, Polish Academy of Sciences, Warszawa, Poland, 2010, pp. 203–223.
40. Szynal, D., Wolynski, W., Goodness-of-fit tests for exponentiality and Rayleigh distribution, *International Journal of Pure and Applied Mathematics*, Vol. 78, No. 5, 2012, pp. 751–772.
41. Nam, K.H., Park, D.H., A study on trend changes for certain parametric families, *Journal of the Korean Society for Quality Management*, Vol. 23, No. 3, 1995, pp. 93–101.
42. Srivastava, A. K., Validation analysis of Dhillon model on different real data sets for reliability modelling, *International Journal of Advance Foundation and Research in Computer (IJAFRC)*, Vol. 1, No. 9, 2014, pp. 18–31.
43. Srivastava, A. K., Kumar, V., A Study of several issues of reliability modelling for a real dataset using different software reliability models, *International Journal of Emerging Technology and Advanced Engineering*, Vol. 5, No. 12, 2015, pp. 49–57.
44. Lim, Y.B. et al., Literature review on the statistical methods in KSQM for 50 years, *Journal of the Korean Society for Quality Management*, Vol. 44, No. 2, 2016, pp. 221–244.

3

Safety, Reliability, Human Factors, and Human Error Basics

3.1 Introduction

The history of the safety field goes back to 1868, when a patent for a barrier safeguard was awarded in the United States [1]. In 1893, U.S. Congress passed the Railway Safety Act. Today, the safety field has branched out into specialized areas such as patient safety, workplace safety, and system safety.

The history of the reliability field may be traced back to the early years of the 1930s, when probability concepts were applied to problems concerning electric power generation [2–4]. However, its real beginning is usually regarded as World War II, when German scientists applied basic reliability concepts for improving the reliability of their V1 and V2 rockets [5]. Today, the reliability field has become a well-developed discipline and has branched out into many specialized areas, including software reliability, mechanical reliability, human reliability, and power system reliability.

The history of human factors can be traced back to 1898 when Frederick W. Taylor conducted a number of studies for determining the most effective designs for shovels [6]. Similarly, the history of human error with regard to engineering systems may directly or indirectly be traced back to 1958 when H.L. Williams pointed out that the reliability of the human element must be considered in the overall prediction of the system reliability; otherwise, the predicted reliability of a system would not depict the actual picture [7]. Today, the field of human error has developed into many areas including health care-related systems, transportation systems, and power systems.

This chapter presents various safety, reliability, human factors, and human error basics considered useful for understanding subsequent chapters of this book.

3.2 Safety Management Principles and Safety and Engineers

There are many safety management principles. The 10 main ones are as follows [8–10]:

- **Principle 1:** The main function of safety is finding and defining the operational errors that result in accidents.
- **Principle 2:** The safety system should be appropriately tailored to fit the company culture effectively.
- **Principle 3:** The causes that lead to unsafe behavior can be highlighted, classified, and controlled.
- **Principle 4:** Safety should be managed just like any other function/activity in an organization. More clearly, management should direct safety by setting attainable safety-related goals and by planning, organizing, and controlling to successfully attain the goals.
- **Principle 5:** The three important symptoms that clearly indicate that something is not right in the management system are an unsafe condition, an unsafe act, and an accident.
- **Principle 6:** In building an effective safety system, the three main subsystems that must be considered carefully are the managerial, the behavioral, and the physical.
- **Principle 7:** Under most circumstances, unsafe behavior is normal behavior because it is the result of normal human beings reacting to the environment surrounding them. Therefore, it is clearly the responsibility of management to conduct appropriate changes to the environment that leads to the unsafe behavior.
- **Principle 8:** There is no single approach/method for achieving safety in an organization/company. However, for a safety system to be effective, it must satisfy certain criteria: have the top level management visibly showing its full support, be quite flexible, involve worker participation, etc.
- **Principle 9:** Management procedures that clearly factor in accountability are the key to successful line safety performance.
- **Principle 10:** There are certain sets of circumstances that can be predicted to lead to severe injuries: high energy sources, certain construction conditions, nonproductive activities, and abnormal, nonroutine tasks.

Today, modern engineering systems/products have become highly sophisticated and complex. Their safety has become a very challenging issue to engineers. Because of global competition and other factors, engineers are pressured to complete new designs rapidly and to do so at lower costs.

In turn, this generally leads to more design-related shortcomings, errors, and causes of accidents.

Design-related shortcomings/deficiencies can directly or indirectly cause accidents or contribute to accidents. Nonetheless, the design shortcoming/deficiency may result because a designer/design [10–20]:

- Failed to foresee unexpected applications of an item/product or all its potential consequences
- Is unfinished, confusing, or incorrect
- Failed to reduce or eliminate the occurrence of human errors
- Violates normal capabilities/tendencies of users
- Relies on product users for avoiding the occurrence of accidents
- Failed to warn properly of potential hazards
- Creates an arrangement of operating controls and other devices that increases reaction time in emergency circumstances or is conducive to the occurrence of human errors
- Incorporates weak warning mechanisms instead of providing a safe design for eradicating potential hazards
- Overlooked providing the appropriate level of protection in a user's personal protective equipment
- Does not appropriately determine or consider the error, action, failure, or omission consequences
- Generates an unsafe characteristic of a product/item
- Failed to prescribe adequate operational procedures in circumstances in which hazards might exist
- Places an unreasonable level of stress on potential product/system operators

3.3 Accident Causation Theories

There are many accident causation theories including the domino theory, the human factors theory, the accident/incident theory, the epidemiological theory, and the combination theory [1]. The first two of these theories are described below, separately.

3.3.1 The Domino Theory

The domino theory is operationalized in 10 statements referred to as the "Axioms of Industrial Safety." These axioms were developed by H.W. Heinrich (an American industrial safety pioneer) and are as follows [1,10,13,14]:

- **Axiom 1:** Most accidents are due to the unsafe acts of humans.
- **Axiom 2:** The reasons why humans commit unsafe acts can be quite useful in selecting appropriate corrective measures.
- **Axiom 3:** Supervisors are the key individuals in the prevention of industrial accidents.
- **Axiom 4:** The most effective accident prevention methods are quite analogous with the productivity and quality approaches.
- **Axiom 5:** Injuries occur from a completed sequence of factors or events, the final one of which is the accident itself.
- **Axiom 6:** There are direct and indirect costs of an accident. Some examples of the direct cost are compensation, liability claims, and medical costs.
- **Axiom 7:** The severity of an injury is largely fortuitous, and the specific accident that caused it is generally preventable.
- **Axiom 8:** An accident can take place only when an individual commits an unsafe act and/or there is some mechanical or physical hazard.
- **Axiom 9:** An unsafe act by an individual or an unsafe condition does not always immediately lead to an accident/injury.
- **Axiom 10:** Management should always assume safety responsibility with absolutely full vigor because they are in the best position for achieving final results.

Furthermore, according to Heinrich, the following five factors in the sequence of events lead to an accident [1,10,13,14]:

- **Factor 1: Ancestry and social environment.** In this case, it is assumed that negative character-related traits, such as recklessness, stubbornness, and avariciousness that might lead persons to behave unsafely, can be acquired as a result of social surroundings or inherited through ancestry.
- **Factor 2: Fault of person.** In this case, it is assumed that negative character-related traits (i.e., whether inherited or acquired) such as nervousness, violent temper, ignorance of safety-related practices, and recklessness, constitute proximate reasons for committing unsafe acts or for the existence of physical or mechanical hazards.
- **Factor 3: Unsafe act/physical or mechanical hazard.** In this case, it is assumed that unsafe acts (such as standing under suspended loads, starting machinery without warning, and removing safeguards) committed by people, and mechanical or physical hazards (such as unguarded gears, absence of rail guards, unguarded point of operation, and inadequate light) are the direct causes for the occurrence of accidents.

- **Factor 4: Accident.** In this case, it is assumed that events such as the striking of humans by flying objects and falls of humans are examples of accidents that result in injury.
- **Factor 5: Injury.** In this case, it is assumed that the injuries directly caused by accidents include fractures and lacerations.

All in all, the central points of the Heinrich theory are the following two items [10,14]:

i. The eradication of the central factor (that is, unsafe act/hazardous condition) definitely negates the action of preceding factors and, in turn, prevents the occurrence of injuries and accidents.
ii. Injuries are the result of the action of all preceding factors.

Additional information on this theory is available in Reference 14.

3.3.2 The Human Factors Theory

The basis for this theory is the assumption that accidents occur due to a chain of events caused by human error. The three main factors that cause the occurrence of human errors are shown in Figure 3.1 [1,10,14].

The factor "inappropriate activities" is concerned with inappropriate activities conducted by an individual due to human error. For example, an individual misjudged the degree of risk involved in a certain task and then conducted the task based on that misjudgment.

The factor "overload" is concerned with an imbalance between the capacity of an individual at any time and the load he/she is carrying in a given state. The capacity of an individual is the product of a number of factors including stress, degree of training, natural ability, physical condition, state of mind, and fatigue. The load carried by an individual is composed of tasks

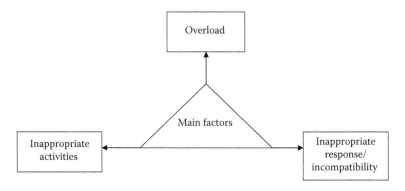

FIGURE 3.1
Human error causing three main factors.

for which he/she has responsibility along with additional burdens due to the following factors [10,14]:

- **Situational factors:** Two examples of these factors are degree of risk and confusing instructions.
- **Internal factors:** Some examples of these factors are personal problems, worry, and emotional stress.
- **Environmental factors:** Two examples of these factors are distractions and noise.

Finally, "inappropriate response/incompatibility" is another main human error causing factor and three examples of inappropriate response by a person are as follows [10,14]:

- **Example 1:** A person completely ignored the recommended safety-associated procedures.
- **Example 2:** A person completely removed a safeguard from a piece of equipment or a machine for improving output.
- **Example 3:** A person detected a hazardous condition but took no proper corrective measure.

Additional information on this theory is available in Reference 14.

3.4 Bathtub Hazard Rate Curve

Usually, a bathtub hazard rate curve is used to describe the failure rate of engineering systems/items and is shown in Figure 3.1. The curve is called the bathtub hazard rate curve because it resembles the shape of a bathtub.

As shown in Figure 3.2, the curve is divided into three regions: burn-in period, useful-life period, and wear-out period. During the burn-in period, the system/item hazard rate decreases with time t, and some of the reasons for the occurrence of failures during this period are as follows [5,15]:

- Poor quality control
- Substandard materials and workmanship
- Inadequate debugging
- Poor manufacturing methods
- Human error
- Poor processes

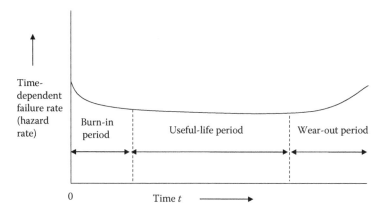

FIGURE 3.2
Bathtub hazard rate curve.

It should be noted that three other terms used for this decreasing hazard rate region are infant mortality region, break-in region, and debugging region. During the useful-life period, the hazard rate remains constant. Some of the reasons for the occurrence of failures in this region are higher random stress than expected, low safety factors, natural failures, abuse, human errors, and undetectable defects.

Finally, during the wear-out period, the hazard rate increases with time t, and some of the reasons for the occurrence of failures in this region are as follows [5,15]:

- Wear due to aging
- Short designed-in life of the system/item under consideration
- Poor maintenance
- Wear due to friction, creep, and corrosion
- Wrong overhaul practices

Mathematically, the following equation can be used to represent the Figure 3.2 bathtub hazard rate curve [16]:

$$\lambda(t) = \alpha\theta(\alpha t)^{\theta-1}e^{(\alpha t)^{\theta}} \tag{3.1}$$

where
$\lambda(t)$ is the time-dependent failure rate (i.e., hazard rate).
t is time.
α is the scale parameter.
θ is the shape parameter.

At $\theta = 0.5$, Equation 3.1 gives the shape of the bathtub hazard rate curve shown in Figure 3.2.

3.5 General Reliability Formulas

There are a number of general formulas used to perform various types of reliability-related analysis. Four of these formulas are presented below.

3.5.1 Failure (or Probability) Density Function

The failure (or probability) density function is defined by [5]

$$f(t) = -\frac{dR(t)}{dt} \qquad (3.2)$$

where
   t is time.
   $R(t)$ is the system/item reliability at time t.
   $f(t)$ is the failure (probability) density function.

EXAMPLE 3.1

The reliability of a system used in a nuclear power plant is expressed by the following equation:

$$R_{ns}(t) = e^{-\lambda_{ns}t} \qquad (3.3)$$

where
   $R_{ns}(t)$ is the nuclear power plant system's reliability at time t.
   λ_{ns} is the nuclear power plant system's constant failure rate.

Obtain an expression for the nuclear power plant system's failure (probability) density function.
By substituting Equation 3.3 into Equation 3.2, we get

$$f(t) = -\frac{de^{-\lambda_{ns}t}}{dt} = \lambda_{ns}e^{-\lambda_{ns}t} \qquad (3.4)$$

Thus, Equation 3.4 is the expression for the nuclear power plant system's failure (probability) density function.

3.5.2 Hazard Rate (Time-Dependent Failure Rate) Function

This is expressed by

$$\lambda(t) = \frac{f(t)}{R(t)} \qquad (3.5)$$

where
   $\lambda(t)$ is the system/item hazard rate (time-dependent failure rate).

By inserting Equation 3.2 into Equation 3.5, we get

$$\lambda(t) = -\frac{1}{R(t)} \cdot \frac{dR(t)}{dt} \tag{3.6}$$

EXAMPLE 3.2

Obtain an expression for the nuclear power plant system's hazard rate with the aid of Equations 3.3 and 3.6 and comment on the final result.
    By inserting Equation 3.3 into Equation 3.6, we obtain

$$\lambda(t) = -\frac{1}{e^{-\lambda_{ns}t}} \cdot \frac{de^{-\lambda_{ns}t}}{dt} = \lambda_{ns} \tag{3.7}$$

Thus, the nuclear power plant system's hazard rate is given by Equation 3.7, and the right-hand side of this equation is not function of time t. Needless to say, λ_{ns} is usually referred to as the constant failure rate of a system/item (in this case, of the nuclear power plant system) because it does not depend on time t.

3.5.3 General Reliability Function

This function can be obtained by using Equation 3.6. Thus, with the aid of Equation 3.6, we obtain

$$-\lambda(t)dt = \frac{1}{R(t)} dR(t) \tag{3.8}$$

By integrating both sides of Equation 3.8 over the time interval $[0,t]$, we et

$$-\int_0^t \lambda(t) \, dt = \int_1^{R(t)} \frac{1}{R(t)} dR(t) \tag{3.9}$$

Since, at $t = 0$, $R(t) = 1$.
Evaluating the right-hand side of Equation 3.9 and rearranging yields

$$\ln R(t) = -\int_0^t \lambda(t) \, dt \tag{3.10}$$

Thus, from Equation 3.10, we get

$$R(t) = e^{\int_0^t \lambda(t)dt} \tag{3.11}$$

Equation 3.11 is the general expression for the reliability function. Thus, it can be used for obtaining the reliability expression of a system/item when its time to failure follow any time-continuous probability distribution (e.g., Weibull, exponential, and Rayleigh).

EXAMPLE 3.3

Assume that the hazard rate of a system used in a nuclear power plant is expressed by Equation 3.1. Obtain an expression for the reliability function of the nuclear power plant system.

By substituting Equation 3.1 into Equation 3.11, we obtain

$$R(t) = e^{-\int_0^t \left\{\alpha\theta(\alpha t)^{\theta-1} e^{(\alpha t)^{\theta}}\right\} dt} = e^{-\left\{e^{(\alpha t)^{\theta}}-1\right\}} \tag{3.12}$$

Thus, Equation 3.12 is the reliability function of the nuclear power plant system.

3.5.4 Mean Time to Failure

The mean time to failure of an item/system can be obtained by using any of the following three formulas [17,18]:

$$MTTF = \int_0^\infty R(t)\, dt \tag{3.13}$$

or

$$MTTF = \lim_{s \to 0} R(s) \tag{3.14}$$

or

$$MTTF = E(t) = \int_0^\infty t f(t)\, dt \tag{3.15}$$

where
$E(t)$ is the expected value.
s is the Laplace transform variable.
$R(s)$ is the Laplace transform of the reliability function $R(t)$.
$MTTF$ is the mean time to failure.

EXAMPLE 3.4

Prove by using Equation 3.3 that Equations 3.13 and 3.14 yield the same result for the nuclear power plant system mean time to failure.

By inserting Equation 3.3 into Equation 3.13, we get

$$MTTF_{ns} = \int_0^\infty e^{-\lambda_{ns}t}\, dt = \frac{1}{\lambda_{ns}} \qquad (3.16)$$

where

$MTTF_{ns}$ is the nuclear power plant system mean time to failure.

By taking the Laplace transform of Equation 3.3, we obtain

$$R_{ns}(s) = \int_0^\infty e^{-st}e^{-\lambda_{ns}t}\, dt = \frac{1}{s + \lambda_{ns}} \qquad (3.17)$$

where

$R_{ns}(s)$ is the Laplace transform of the nuclear power plant system's reliability function $R_{ns}(t)$.

Substituting Equation 3.17 into Equation 3.14 yields

$$MTTF_{ns} = \lim_{s \to 0} \frac{1}{(s + \lambda_{ns})} = \frac{1}{\lambda_{ns}} \qquad (3.18)$$

As Equations 3.16 and 3.18 are identical, it proves that Equations 3.13 and 3.14 yield the same result for the nuclear power plant system mean time to failure.

3.6 Reliability Networks

A system in the area of nuclear power generation can form various configurations in performing reliability analysis. Thus, this section is concerned with the reliability evaluation of such commonly occurring configurations or networks.

3.6.1 Series Network

This network is the simplest reliability network/configuration, and its block diagram is shown in Figure 3.3. The diagram denotes a k-unit system, and

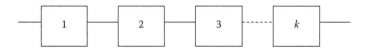

FIGURE 3.3
A k-unit series system/network/configuration.

each block in the diagram represents a unit/part. For the successful operation of the series system/network/configuration, all k units must operate normally. In other words, if any one of the k units fails, the series system/network/configuration fails.

The series system, shown in Figure 3.3, reliability is expressed by

$$R_s = P(E_1 E_2 E_3 \ldots E_k) \tag{3.19}$$

where
E_j is the successful operation (i.e., success event) of unit j, for $j = 1, 2, 3, \ldots, k$.
$P(E_1 E_2 E_3 \ldots E_k)$ is the occurrence probability of events E_1, E_2, E_3, ..., E_k.
R_s is the series system reliability.

For independently failing units, Equation 3.19 becomes

$$R_s = P(E_1)P(E_2)P(E_3)\ldots P(E_k) \tag{3.20}$$

where
$P(E_j)$ is the occurrence probability of event E_j, for $j = 1, 2, 3, \ldots, k$.

If we let $R_j = P(E_j)$, for $j = 1, 2, 3, \ldots, k$, Equation 3.20 becomes

$$R_s = R_1 R_2 R_3 \ldots R_k = \prod_{j=1}^{k} R_j \tag{3.21}$$

where
R_j is the unit j reliability for $j = 1, 2, 3, \ldots, k$.

For constant failure rate λ_j (i.e., $\lambda_j(t) = \lambda_j$) of unit j from Equation 3.11, we get

$$R_j(t) = e^{-\lambda_j t} \tag{3.22}$$

where
$R_j(t)$ is the unit j reliability at time t.

By substituting Equation 3.22 into Equation 3.21, we obtain

$$R_s(t) = e^{-\sum_{j=1}^{k} \lambda_j t} \tag{3.23}$$

where
$R_s(t)$ is the series system reliability at time t.

By substituting Equation 3.23 into Equation 3.13, we obtain the following expression for the series system mean time to failure:

$$MTTF_s = \int_0^\infty e^{-\sum_{j=1}^{k} \lambda_j t}\, dt = \frac{1}{\sum_{j=1}^{k} \lambda_j} \tag{3.24}$$

where
$MTTF_s$ is the series system mean time to failure.

By inserting Equation 3.23 into Equation 3.6, we obtain the following expression for the series system hazard rate:

$$\lambda_s(t) = -\frac{1}{e^{-\sum_{j=1}^{k} \lambda_j t}} \left[-\sum_{j=1}^{k} \lambda_j \right] e^{-\sum_{j=1}^{k} \lambda_j t} = \sum_{j=1}^{k} \lambda_j \tag{3.25}$$

where
$\lambda_s(t)$ is the series system hazard rate.

It should be noted that the right-hand side of Equation 3.25 is independent of time t. Thus, the left-hand side of this equation is simply λ_s, the series system failure rate. It means that whenever we add up failure rates of items/units, we automatically assume that these items/units form a series network/configuration, a worst-case design scenario with regard to reliability.

EXAMPLE 3.5

Assume that a system used in a nuclear power plant is composed of five independent and identical units, and the constant failure rate of each unit is 0.0005 failures per hour. All five units must operate normally for the system to function successfully. Calculate the system's reliability for a 10-hour mission, mean time to failure, and failure rate.

By inserting the specified data values into Equation 3.23, we obtain

$$R_s(10) = e^{-(0.0005)(5)(10)} = 0.9753$$

Substituting the specified data values into Equation 3.24 yields

$$MTTF_s = \frac{1}{5(0.0005)} = 400 \text{ hours}$$

By substituting the given data values into Equation 3.25, we get

$$\lambda_s = 5(0.0005) = 0.0025 \text{ failures/hour}$$

Thus, the nuclear power plant system's reliability, mean time to failure, and failure rate are 0.9753, 400 hours, and 0.0025 failures/hour, respectively.

3.6.2 Parallel Network

This network represents a system with k units/items operating simultaneously. For the successful operation of the system, at least one of these units/items must operate normally. The block diagram of a k-unit parallel system is shown in Figure 3.4, and each block in the diagram denotes a unit.

The failure probability of the parallel system shown in Figure 3.4 is expressed by

$$F_p = P\left(\bar{x}_1 \bar{x}_2 \bar{x}_3 \ldots \bar{x}_k\right) \tag{3.26}$$

where

F_p is the failure probability of the parallel system.
$\bar{x}_j$ is the failure (i.e., failure event) of unit j; for $j = 1, 2, 3, \ldots, k$.
$P(\bar{x}_1 \bar{x}_2 \bar{x}_3 \ldots \bar{x}_k)$ is the probability of occurrence of events $\bar{x}_1, \bar{x}_2, \bar{x}_3, \ldots,$ and $\bar{x}_k$.

For independently failing parallel units, Equation 3.26 is written as

$$F_p = P(\bar{x}_1)P(\bar{x}_2)P(\bar{x}_3)\ldots P(\bar{x}_k) \tag{3.27}$$

where

$P(\bar{x}_j)$ is the occurrence probability of failure event $\bar{x}_j$, for $j = 1, 2, 3, \ldots, k$.

If we let $F_j = P(\bar{x}_j)$, for $j = 1, 2, 3, \ldots, k$, then Equation 3.27 becomes

$$F_p = F_1 F_2 F_3 \ldots F_k = \prod_{j=1}^{k} F_j \tag{3.28}$$

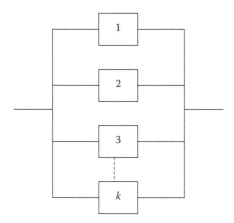

FIGURE 3.4
Block diagram of a parallel system/network with k units.

where
  F_j is the unit j failure probability, for $j = 1, 2, 3, \ldots, k$.

By subtracting Equation 3.28 from unity, we obtain

$$R_p = 1 - F_p = 1 - \prod_{j=1}^{k} F_j \tag{3.29}$$

where
  R_p is the parallel system/network reliability.

For constant failure rate λ_j of unit j, subtracting Equation 3.22 from unity and then substituting it into Equation 3.29 yields

$$R_p(t) = 1 - \prod_{j=1}^{k} \left(1 - e^{-\lambda_j t}\right) \tag{3.30}$$

where
  $R_p(t)$ is the parallel system/network reliability at time t.

For identical units, Equation 3.30 becomes

$$R_p(t) = 1 - (1 - e^{-\lambda t})^k \tag{3.31}$$

where
  λ is the unit constant failure rate.

By inserting Equation 3.31 into Equation 3.13, we obtain the following expression for the parallel system/network mean time to failure:

$$MTTF_p = \int_0^\infty [1-(1-e^{-\lambda t})^k]\,dt = \frac{1}{\lambda}\sum_{j=1}^k \frac{1}{j} \tag{3.32}$$

where
$MTTF_p$ is the identical unit parallel system/network mean time to failure.

EXAMPLE 3.6

Assume that a system used in a nuclear power plant is composed of two identical, independent, and active units. At least one of the units must operate normally for the system to operate successfully. The constant failure rate of a unit is 0.001 failures per hour.

Calculate the system reliability for a 400-hour mission and mean time to failure.

By substituting the specified data values into Equation 3.31, we get:

$$R_p(400) = 1-[1-e^{-(0.001)(400)}]^2 = 0.8913$$

Inserting the given data values into Equation 3.32 yields

$$MTTF_p = \frac{1}{(0.001)}\left[1+\frac{1}{2}\right] = 1500 \text{ hours}$$

Thus, the system reliability and mean time to failure are 0.8913 and 1500 hours, respectively.

3.6.3 *k*-out-of-*n* Network

In this case, the system/network is composed of a total of n active units, and at least k units out of n active units must function normally for the successful operation of the system/network. The block diagram of a k-out-of-n unit system/network is shown in Figure 3.5, and each block in the diagram represents a unit. The series and parallel networks are special cases of this network for $k = n$ and $k = 1$, respectively.

By using the binomial distribution for independent and identical units, we write the following equation for reliability of k-out-of-n unit network shown in Figure 3.5:

$$R_{k/n} = \sum_{i=k}^n \binom{n}{i} R^i (1-R)^{n-i} \tag{3.33}$$

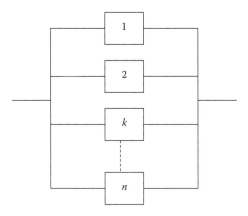

FIGURE 3.5
k-out-of-*n* unit network block diagram.

where

$$\binom{n}{i} = \frac{n!}{(n-i)!i!} \tag{3.34}$$

$R_{k/n}$ is the *k*-out-of-*n* network/system reliability.
R is the unit reliability.

For constant failure rates of the identical units, using Equations 3.11 and 3.33, we obtain

$$R_{k/n}(t) = \sum_{i=k}^{n} \binom{n}{i} e^{-i\lambda t}(1 - e^{-\lambda t})^{n-i} \tag{3.35}$$

where
  λ is the unit constant failure rate.
  $R_{k/n}(t)$ is the *k*-out-of-*n* network/system reliability at time *t*.

By substituting Equation 3.35 into Equation 3.13, we get

$$MTTF_{k/n} = \int_{0}^{\infty} \left[\sum_{i=k}^{n} \binom{n}{i} e^{-i\lambda t}(1 - e^{-\lambda t})^{n-i} \right] dt = \frac{1}{\lambda} \sum_{i=k}^{n} \frac{1}{i} \tag{3.36}$$

where
  $MTTF_{k/n}$ is the *k*-out-of-*n* network/system mean time to failure.

EXAMPLE 3.7

Assume that a system used in a nuclear power plant has three identical, independent, and active units in parallel. At least two units must function normally for the successful operation of the system. Calculate the system mean time to failure if the unit constant failure rate is 0.0005 failures per hour.

By inserting the specified data values into Equation 3.36, we obtain

$$MTTF_{2/3} = \frac{1}{(0.0005)} \left[\frac{1}{2} + \frac{1}{3} \right] = 1666.67 \text{ hours}$$

Thus, the system mean time to failure is 1666.67 hours.

3.6.4 Standby System

This is another network/configuration/system in which only one unit operates and k units are kept in their standby mode. The system contains $(k + 1)$ units, and as soon as the functioning/operating unit fails, the switching mechanism detects the failure and turns on one of the standby units. The system fails when all the standby units malfunction/fail.

The block diagram of a standby system with one operating/functioning and k standby units is shown in Figure 3.6. Each block in the diagram denotes a unit.

With the aid of Figure 3.6 diagram for independent and identical units, perfect switching mechanism and standby units, and time-dependent unit failure rate, we write the following equation for the standby system reliability [19]:

$$R_{ss}(t) = \sum_{j=0}^{k} \frac{\left[\left[\int_{0}^{t} \lambda(t)dt \right]^{j} e^{-\int_{0}^{t} \lambda(t)dt} \right]}{j!} \tag{3.37}$$

where
 $R_{ss}(t)$ is the standby system reliability at time t.
 $\lambda(t)$ is the unit time-dependent failure rate or hazard rate.

For constant unit failure rate (i.e., $\lambda(t) = \lambda$), Equation 3.37 yields

$$R_{ss}(t) = \frac{\left[\sum_{j=0}^{k} (\lambda t)^{j} e^{-\lambda t} \right]}{j!} \tag{3.38}$$

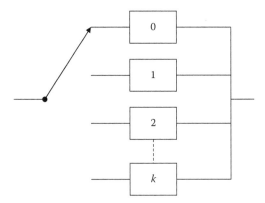

FIGURE 3.6
Block diagram of a standby system with one operating and k standby units.

where
  λ is the unit constant failure rate.
  By inserting Equation 3.38 into Equation 3.13, we get

$$MTTF_{ss} = \int_0^\infty \left[\sum_{j=0}^k \frac{(\lambda t)^j e^{-\lambda t}}{j!} \right] dt = \frac{k+1}{\lambda} \qquad (3.39)$$

where
  $MTTF_{ss}$ is the standby system mean time to failure.

EXAMPLE 3.8

A standby system used in a nuclear power plant is composed of two identical and independent units: one operating, the other on standby. The unit constant failure rate is 0.008 failures per hour. Calculate the standby system reliability for a 200-hour mission and mean time to failure if the switching mechanism is perfect and the standby unit remains as good as new in its standby mode.
  By substituting the given data values into Equation 3.38, we get

$$R_{ss}(200) = \sum_{j=0}^1 \frac{\left[(0.008)(200)\right]^j e^{-(0.008)(200)}}{j!} = 0.5249$$

Similarly, by inserting the given data values into Equation 3.39, we get

$$MTTF_{ss} = \frac{2}{0.008} = 250 \text{ hours}$$

Thus, the standby system reliability and mean time to failure are 0.5249 and 250 hours, respectively.

3.6.5 Bridge Network

Sometimes systems used in nuclear power plants may form a bridge network, as shown in Figure 3.7. Each block in the figure denotes a unit, and all units are labeled with numerals.

For independently failing units in the bridge network shown in Figure 3.7, the network reliability is expressed by [20]:

$$R_b = 2R_1R_2R_3R_4R_5 + R_1R_3R_5 + R_2R_3R_4 + R_2R_5 + R_1R_4$$
$$- R_1R_2R_3R_4 - R_1R_2R_3R_5 - R_2R_3R_4R_5 - R_1R_2R_4R_5 - R_3R_4R_5R_1 \qquad (3.40)$$

where
R_i is the unit i reliability, for $i = 1, 2, 3, 4, 5$.
R_b is the bridge network reliability.

For identical units, Equation 3.40 simplifies to

$$R_b = 2R^5 - 5R^4 + 2R^3 + 2R_2 \qquad (3.41)$$

where
R is the unit reliability.

For constant unit failure rate, with the aid of Equations 3.11 and 3.41, we obtain

$$R_b(t) = 2e^{-5\lambda t} - 5e^{-4\lambda t} + 2e^{-3\lambda t} + 2e^{-2\lambda t} \qquad (3.42)$$

where
$R_b(t)$ is the bridge network reliability at time t.
λ is the unit constant failure rate.

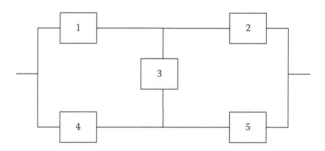

FIGURE 3.7
A five nonidentical unit bridge network.

By substituting Equation 3.42 into Equation 3.13, we obtain

$$MTTF_b = \int_0^\infty (2e^{-5\lambda t} - 5e^{-4\lambda t} + 2e^{-3\lambda t} + 2e^{-2\lambda t})dt = \frac{49}{60\lambda} \qquad (3.43)$$

where
$MTTF_b$ is the bridge network mean time to failure.

EXAMPLE 3.9

Assume that a nuclear plant system has five identical and independent units forming a bridge network. The constant failure rate of each unit is 0.0002 failures per hour.

Calculate the bridge network reliability for a 250-hour mission and mean time to failure.

By substituting the specified data values into Equation 3.42, we get

$$R_b(250) = 2e^{-5(0.0002)(250)} - 5e^{-4(0.0002)(250)} + 2e^{-3(0.0002)(250)} + 2e^{-2(0.0002)(250)}$$
$$= 0.9950$$

Similarly, inserting the given data value into Equation 3.43 yields

$$MTTF_b = \frac{49}{60(0.0002)} = 4083.33 \text{ hours}$$

Thus, the bridge network reliability and mean time to failure are 0.9950 and 4083.33 hours, respectively.

3.7 Human Factor Objectives and Typical Human Behaviors

There are many human factors objectives. They may be grouped under the following four categories [10,21]:

- **Category 1: Objectives affecting reliability and maintainability.** These objectives are concerned with items such as improving reliability and maintainability, reducing training-related requirements, and reducing the manpower need.
- **Category 2: Fundamental operational objectives.** These objectives are basically concerned with items such as reducing the occurrence of human errors, improving system performance, and improving safety.

- **Category 3: Objectives affecting operators and users.** These objectives are concerned with items such as reducing boredom, fatigue, physical stress, and monotony as well as increasing user acceptance and ease of use, improving the work environment, and increasing aesthetic appearance.
- **Category 4: Miscellaneous objectives.** These objectives are concerned with items such as reducing equipment and time losses and increasing production economy.

Over the years, professionals working in the area of human factors have highlighted many typical human behaviors. Eight of these behaviors are as follows [10,22]:

   i. Quite often humans tend to hurry.

   ii. Humans have become accustomed to certain color meanings.

   iii. Humans usually regard a manufactured item/product as being safe.

   iv. The attention of humans is drawn to factors such as loud noises, bright and vivid colors, flashing lights, and bright lights.

   v. Humans get quite easily confused with unfamiliar things.

   vi. Humans generally expect to turn on the electrical power (the switches have to move upward, or to the right, etc.).

   vii. Humans expect that faucets and valve handles will rotate counterclockwise for increasing the flow of steam, liquid, or gas.

   viii. Humans quite often use their hands first for testing or exploring the unknown.

3.8 Human Sensory Capacities

Humans possess many useful sensors: hearing, taste, sight, smell, and touch. More specifically, humans can sense vibration, temperature, acceleration (shock), pressure, linear motion, position, and rotation. It simply means that a clear understanding of human sensory capacities can be very useful in reducing the occurrence of human errors in nuclear power plants. Thus, some of the human sensory capacities are described below [10,23].

- **Sensory capacity: Touch.** Touch is closely related to the ability of humans to interpret visual and auditory stimuli. The sensory cues received by the skin and muscles can be used for sending messages to the brain, thus relieving the human ears and eyes of the workload to a certain degree.

Additional information on this topic is available in Reference 24.

- **Sensory capacity: Sight.** This is stimulated by the electromagnetic radiation of certain wavelengths, often referred to as the visible segment of the electromagnetic spectrum. The various areas of the spectrum, as seen by eyes, appear to vary in degree of brightness. For example, during the day, the human eyes are quite sensitive to greenish-yellow light with a wavelength of approximately 5500 Å units [23]. Furthermore, the eyes see differently from different angles.

  Moreover, generally the eyes perceive all colors when they are looking straight ahead; however, the perception of color reduces with the increment in the viewing angle.

- **Sensory capacity: Vibration.** Past experiences over the years indicate that the existence of vibration could be quite detrimental to the performance of mental and physical tasks by people. There are a number of vibration parameters: amplitude, frequency, velocity, acceleration, and jolt. More specifically, a large amplitude and low-frequency vibrations contribute to eye strain, headaches, deterioration in ability to read and interpret instruments, motion sickness, and fatigue [23]. In addition, low amplitude and high-frequency vibrations can be quite fatiguing.

- **Sensory capacity: Noise.** It may simply be stated as sounds that lack coherence. Humans' reaction to the noise problem extends beyond the auditory system. It can lead to feelings such as fatigue, boredom, or irritability. Excessive noise can result in various problems including reduction in a worker's efficiency, adverse effects on tasks requiring a high degree of muscular coordination or intense concentration, and loss in hearing if exposed for long periods.

3.9 Useful Human Factor-Related Guidelines

Over the years, professionals working in the area of human factors have developed many human factor-related guidelines considered quite useful for application in designing engineering systems. Nine of these guidelines are as follows [10,23,24]:

- **Guideline 1:** Develop a human factor-related checklist for application in design and production phases.

- **Guideline 2:** Review system objectives with respect to human factors.
- **Guideline 3:** Acquire all applicable human factor-related design reference documents.
- **Guideline 4:** Use human factors specialist services as considered appropriate.
- **Guideline 5:** Review with care final production drawings with regard to human factors.
- **Guideline 6:** Conduct experiments when cited reference guides fail to provide appropriate information for design-related decision.
- **Guideline 7:** Use mock-ups to "test" the effectiveness of all user-hardware interface designs.
- **Guideline 8:** Conduct appropriate field tests of the system design before approving it for delivery to users/customers.
- **Guideline 9:** Fabricate a hardware prototype (if possible) and evaluate it under applicable environments.

3.10 Useful Mathematical Human Factor-Related Formulas

Over the years, various types of mathematical formulas for estimating human factor-related information have been developed. Four of these formulas considered quite useful to study direct or indirect human error in systems used in nuclear power plants are presented below.

3.10.1 Formula I: Inspector Performance

This formula is concerned with estimating inspector performance with respect to inspection-oriented tasks. Thus, the inspector performance is expressed by [10,25]:

$$\theta = \frac{T_r}{\alpha - \beta} \tag{3.44}$$

where
θ is the inspector performance expressed in minutes per correct inspection.
α is the number of patterns inspected.
β is the number of inspector errors.
T_r is the reaction time expressed in minutes.

Additional information on this formula is available in Reference 25.

3.10.2 Formula II: Character Height

This formula is concerned with estimating the character height at the viewing distance of 28 inches, as generally the instrument panels are located at a viewing distance of 28 inches for the comfortable performance as well as control of adjustment-oriented tasks. Thus, the character height is expressed by [26,27]:

$$H_c = \frac{H_{sc}VD_r}{28} \qquad (3.45)$$

where
  H_c is the character height at the required viewing distance, expressed in inches.
  H_{sc} is the standard character height at a viewing distance of 28 inches.
  VD_r is the required viewing distance expressed in inches.

EXAMPLE 3.10

Assume that a systems operator in a nuclear power plant has to read a meter from a distance of 50 inches and the standard character height at a viewing of 28 inches at low luminance is 0.40 inches. Calculate the height of numerals for the stated viewing distance.
  By inserting the given data values into Equation 3.45, we get:

$$H_c = \frac{(0.40)(50)}{28} = 0.71 \text{ inches}$$

Thus, the height of numerals for the stated viewing distance is 0.71 inches.

3.10.3 Formula III: Rest Period

This formula is concerned with estimating the length of the rest period needed for humans conducting various types of tasks. The length of the needed rest is expressed by [10,28,29]:

$$RT_r = \frac{T_w(E_a - C_s)}{(E_a - \gamma)} \qquad (3.46)$$

where
  RT_r is the required rest time expressed in minutes.
  E_a is the average energy expenditure/cost expressed in kilocalories per minute of work.
  γ is the approximate resting level expressed in kilocalories per minute (generally, the value of γ is taken as 1.5)
  C_s is kilocalories per minute adopted as standard.
  T_w is the working time expressed in minutes.

Example 3.11

Assume that a maintenance worker in a nuclear power plant is conducting a maintenance task for 90 minutes, and his/her average energy expenditure is 6 kilocalories per minute. Calculate the length of the required rest time for the worker if $C_s = 3$ kilocalories per minute.

By inserting the specified data values into Equation 3.46, we obtain:

$$RT_r = \frac{(90)(6-3)}{(6-1.5)} = 60 \text{ minutes}$$

Thus, the length of the required rest time is 60 minutes.

3.10.4 Formula IV: Glare Constant

This formula is concerned with estimating the glare constant value as human errors can occur (in performing various operation and maintenance tasks in nuclear power plants) due to glare. The glare constant is expressed by [10,29]:

$$\alpha = \frac{(\gamma^{1.6})(\beta^{0.8})}{(BL)\lambda^2} \tag{3.47}$$

where
α is the glare constant.
BL is the general background luminance.
γ is the source luminance.
β is the solid angle subtended at the eye by the source.
λ is the angle between the direction of the glare source and the viewing direction.

Additional information on this formula is available in Reference 29.

3.11 Reasons for Human Error Occurrence and Types of Human Errors

There are many reasons for the occurrence of human errors in engineering systems. Some of the main ones are as follows [10,30,31]:

- Poor equipment/system design
- Inadequate or poorly written equipment/system operating and maintenance procedures

- Complex tasks
- Inadequate work tools
- Poor job environment (i.e., poor lighting, high/low temperature, high noise level, crowded work space, etc)
- Poor work layout
- Poor training or skill of concerned personnel
- Poor motivation of involved personnel

There are many types of human errors. They may be grouped under the following seven types: [10,30–32]:

- **Design errors:** These errors are the result of poor design. The causes for the occurrence of these errors include failure to implement human needs in the design, failure to ensure the man-machine interaction effectiveness, and assigning inappropriate functions to humans.

  An example of design errors is the placement of controls and displays so far apart that operators find it difficult to use them effectively.

- **Maintenance errors:** These errors occur in the field environment due to oversights by the maintenance personnel. As the equipment becomes old, the chances for the occurrence of such errors may increase because of the increase in the frequency of maintenance.

  Some examples of maintenance errors are calibrating equipment incorrectly, applying the wrong grease at appropriate points of equipment, and repairing the failed equipment incorrectly.

- **Assembly errors:** These errors occur during the product assembly process due to humans. They occur due to causes such as inadequate illumination, poor blueprints, poorly designed work layout, excessive temperature in the work area, excessive noise level, and poor communication of related information.

- **Installation errors:** These errors occur due to various reasons including using the wrong installation blueprints/instructions or failing to install equipment according to the specifications of the manufacturer.

- **Operator errors:** These errors are the result of operator mistakes. The conditions that lead to operator errors include complex tasks, lack of proper procedures, operator carelessness, poor environments, and poor personnel selection and training.

- **Handling errors:** These errors basically occur because of poor transportation or storage facilities. More specifically, such facilities are not as specified by the manufacturers of equipment.

- **Inspection errors:** These errors occur because of less than 100% accuracy of inspectors. As per Reference 33, an average inspection effectiveness is about 85%. An example of an inspection error is accepting and rejecting out-of-tolerance and in-tolerance parts, respectively.

3.12 Problems

1. What are the 10 main principles of safety management?
2. Describe the human factors theory.
3. Describe the bathtub hazard rate curve.
4. Write the general formulas for the following three functions:
   a. Failure (or probability) density function
   b. Hazard rate (time-dependent failure rate) function
   c. General reliability function
5. Assume that a system used in a nuclear power plant is composed of seven identical and independent units, and the constant failure rate of each unit is 0.0002 failures per hour. All seven units must operate normally for the system to operate successfully. Calculate the system's reliability for a 20-hour mission, mean time to failure, and failure rate.
6. Assume that a system used in a nuclear power plant has four independent, identical, and active units in parallel. At least two units must operate normally for the successful operation of the system. Calculate the system mean time to failure if the unit constant failure rate is 0.0001 failures per hour.
7. Assume that a nuclear power plant system has five identical and independent units forming a bridge network. The constant failure rate of each unit is 0.0005 failures per hour.

   Calculate the bridge network reliability for an 80-hour mission and mean time to failure.
8. Describe at least eight typical human behaviors.
9. Describe the following four types of human errors:
   a. Operator errors
   b. Design errors
   c. Maintenance errors
   d. Inspection errors
10. Describe the following three human sensory capacities:
   a. Sight
   b. Touch
   c. Vibration

References

1. Goetsch, D.L., *Occupational Safety and Health*, Prentice Hall, Englewood Cliffs, New Jersey, 1996.
2. Layman, W.J., Fundamental consideration in preparing a master plan, *Electrical World*, Vol. 101, 1933, pp. 778–792.
3. Smith, S.A., Service reliability measured by probabilities of outage, *Electrical World*, Vol. 103, 1934, pp. 371–374.
4. Dhillon, B.S., *Power System Reliability, Safety, and Management*, Ann Arbor Science Publishers, Ann Arbor, Michigan, 1983.
5. Dhillon, B.S., *Design Reliability: Fundamentals and Applications*, CRC Press, Boca Raton, Florida, 1999.
6. Chapanis, A., *Man-Machine Engineering*, Wadsworth Publishing Company, Belmont, California, 1965.
7. Williams, H.L., *Reliability Evaluation of the Human Component in Man-Machine Systems*, Electrical Manufacturing, April 1958, pp. 78–82.
8. Petersen, D., *Techniques of Safety Management*, McGraw-Hill, New York, 1971.
9. Petersen, D., *Safety Management*, American Society of Safety Engineers, Des Plaines, Illinois, 1998.
10. Dhillon, B.S., *Safety and Human Error in Engineering Systems*, CRC Press, Boca Raton, Florida, 2013.
11. Dhillon, B.S., *Transportation Systems Reliability and Safety*, CRC Press, Boca Raton, Florida, 2011.
12. Hammer, W., Price, D., *Occupational Safety Management and Engineering*, Prentice Hall, Upper Saddle River, New Jersey, 2001.
13. Heinrich, H.W., *Industrial Accident Prevention*, McGraw-Hill, New York, 1959.
14. Heinrich, H.W., Petersen, D., Roos, N., *Industrial Accident Prevention*, McGraw-Hill, New York, 1980.
15. Kapur, K.C., Reliability and maintainability, in *Handbook of Industrial Engineering*, edited by G. Salvendy, John Wiley & Sons, New York, 1982, pp. 8.5.1–8.5.34.
16. Dhillon, B.S., Life distributions, *IEEE Transactions on Reliability*, Vol. 30, No. 5, 1981, pp. 457–460.
17. Shooman, M.L., *Probabilistic Reliability: An Engineering Approach*, McGraw-Hill, New York, 1968.
18. Dhillon, B.S., *Reliability, Quality, and Safety for Engineers*, CRC Press, Boca Raton, Florida, 2005.
19. Sandler, G.H., *System Reliability Engineering*, Prentice Hall, Englewood Cliffs, New Jersey, 1963.
20. Lipp, J.P., Topology of switching elements versus reliability, *Transactions on IRE Reliability and Quality Control*, Vol. 7, 1957, pp. 21–34.
21. Chapanis, A., *Human Factors in Systems Engineering*, John Wiley & Sons, New York, 1996.
22. Woodson, W.E., *Human Factors Design Handbook*, McGraw-Hill, New York, 1981.
23. AMCP-706-134, *Engineering Design Handbook: Maintainability Guide for Design*, prepared by the U.S. Army Material Command, Alexandria, Virginia, 1972.

24. Dhillon, B.S., *Advanced Design Concepts for Engineers*, Technomic Publishing Company, Lancaster, Pennsylvania, 1998.
25. Drury, C.G., Fox, J.G., eds., *Human Reliability in Quality Control*, John Wiley & Sons, New York, 1975.
26. Huchingson, R.D., *New Horizons for Human Factors in Design*, McGraw-Hill, New York, 1982.
27. McCormick, E.J., Sanders, M.S., *Human Factors in Engineering and Design*, McGraw-Hill, New York, 1982.
28. Murrell, K.F.H., *Human Performance in Industry*, Reinhold Publishing Company, New York, 1965.
29. Oborne, D.J., *Ergonomics at Work*, John Wiley & Sons, New York, 1982.
30. Meister, D., The Problem of Human-Initiated Failures, *Proceedings of the Eighth National Symposium on Reliability and Quality Control*, 1962, pp. 234–239.
31. Dhillon, B.S., *Human Reliability: With Human Factors*, Pergamon Press, New York, 1986.
32. Cooper, J.L., Human initiated failures and man-function reporting, *IRE Trans. Human Factors*, Vol. 10, 1961, pp. 104–109.
33. McCornack, R.L., *Inspector Accuracy: A Study of the Literature*, Report No. SCTM 53–61 (14), Sandia Corporation, Albuquerque, New Mexico, 1961.

4

Methods for Performing Safety, Reliability, Human Factors, and Human Error Analysis in Nuclear Power Plants

4.1 Introduction

Over the years, a large number of publications in the areas of safety, reliability, human factors, and human error have appeared in the form of journal articles, conference proceedings articles, technical reports, and books [1–5]. Many of these publications report the development of various types of methods and approaches for performing safety, reliability, human factors, and human error analyses. Some of these methods can be used for performing safety, reliability, human factors, and human error analysis in nuclear power plants. The others are more confined to a specific field (i.e., safety, reliability, human factors, or human error).

Two examples of these methods and approaches that can be used for performing safety, reliability, and human error analysis in nuclear power plants are fault tree analysis (FTA) and failure modes and effect analysis (FMEA). FTA was developed in the early 1960s at Bell Telephone Laboratories for performing safety and reliability-related analysis of the Minuteman Launch Control System [6,7]. Nowadays, FTA is being utilized to analyze various types of problems in many diverse areas including engineering, health care, and management [1,2,7–9].

FMEA was developed in the early 1950s by the U.S. Department of Defense to perform analysis of engineering systems from the reliability aspect. Nowadays, FMEA is being utilized to analyze various types of problems in a wide range of areas including safety, reliability, health care, and human factors [1–3,7,8].

This chapter presents a number of methods and approaches considered useful for performing safety, reliability, human factors, and human error analysis in nuclear power plants, extracted from the published literature in the areas of safety, reliability, and human factors.

4.2 Technique of Operations Review (TOR)

This method seeks to highlight systemic causes rather than assigning blame with respect to safety. It was developed in the early 1970s by D.A. Weaver of the American Society of Safety Engineers [1,10,11]. Technique of operations review (TOR) may simply be described as a hands-on analytical approach to identify the root system causes of an operation failure. It allows management and employees to work jointly in conducting analyses of workplace related incidents, accidents, and failures.

TOR makes use of a worksheet containing simple terms, basically, requiring "yes/no" decisions. An incident occurring at a certain location and time involving certain persons activates TOR. Furthermore, it should be noted that because TOR is not a hypothetical process, it demands systematic evaluation of the actual circumstances surrounding the incident.

TOR is composed of the following steps [1,10,11]:

- **Step A:** Form the TOR team by carefully selecting its members from all concerned areas.
- **Step B:** Hold a roundtable meeting for disseminating common knowledge to all members of the team.
- **Step C:** Highlight one key systemic factor that directly or indirectly played an instrumental role in the incident/accident occurrence. This factor serves as a starting point for further investigation into the occurrence of the incident/accident, and it must be based on the team consensus.
- **Step D:** Make use of team consensus in responding to a sequence of "yes/no" options.
- **Step E:** Evaluate all the highlighted factors by ensuring the clear existence of team consensus with respect to the evaluation of each factor.
- **Step F:** Prioritize all the contributory factors by starting with the most serious one.
- **Step G:** Develop appropriate preventive/corrective strategies with respect to each and every contributory factor.
- **Step H:** Implement all the strategies.

Finally, it should be noted that the main strength of this method (i.e., TOR) is the involvement of line personnel in the analysis, and its main weakness is an after-the-fact process.

4.3 Root Cause Analysis (RCA)

This method was developed by the U.S. Department of Energy for investigating industrial-related incidents [12–14]. Root cause analysis (RCA) may simply be described as a systemic investigation approach that uses information collected during the assessment of an accident for determining the underlying factors for shortcomings that caused the accident [12–14].

The following steps are involved in performing RCA [14,15]:

- **Step A:** Educate personnel involved in RCA.
- **Step B:** Inform appropriate staff personnel when a sentinel event is reported.
- **Step C:** Form an RCA team made up of appropriate individuals.
- **Step D:** Appropriately prepare for and hold the first team meeting.
- **Step E:** Determine the event sequence.
- **Step F:** Separate and highlight each event sequence that may have directly or indirectly been a contributory factor in the sentinel event occurrence.
- **Step G:** Brainstorm regarding the factors surrounding the selected events that may have directly or indirectly been a contributory factor to the sentinel event occurrence.
- **Step H:** Affinitize with the findings of the brainstorm session.
- **Step I:** Develop the action plan.
- **Step J:** Distribute the action plan and the RCA document to all involved individuals.

Over the years, RCA has been used in many areas, and some of its benefits and drawbacks observed are as follows [14,16]:

Benefits

- It is a quite effective tool for highlighting and addressing systems and organizational-related issues.
- It is a quite well-structured and process-focused approach.
- The systematic application of the method can highlight common root causes that directly or indirectly link a disparate collection of accidents.

Drawbacks

- It is a quite time-consuming and labor-intensive approach.
- It is quite possible to be tainted by hindsight bias.

- It is impossible for determining if the root cause established by the analysis is really the actual cause for the occurrence of the accident.
- In essence, it (i.e., RCA) is basically an uncontrolled case study.

4.4 Interface Safety Analysis (ISA)

This method is concerned with determining the incompatibilities between assemblies and subsystems of an item/product that could lead to accidents. The analysis establishes that completely distinct units/parts can be integrated into a quite viable system, and an individual unit's or part's normal operation will not deteriorate the performance or damage another unit/part or the whole system/product. Although interface safety analysis (ISA) considers various relationships, they can be grouped under three classifications, as shown in Figure 4.1 [14,17].

The physical relationships are connected to the units' or items' physical aspects. For example, two units or items might be quite well designed and manufactured and function well individually, but they may fail to fit together because of dimensional differences, or they present other problems that may result in safety-related issues. Three typical examples of the other problems are as follows [14,17]:

- **Example 1:** Very little clearance between units/items; thus, the units/items may be damaged during removal or replacement process
- **Example 2:** Restricted or impossible access to or egress from equipment
- **Example 3:** Impossible to join, tighten, or mate components/parts properly

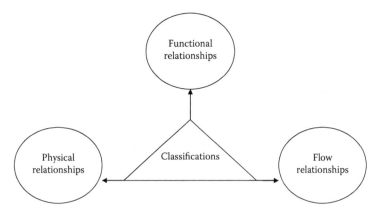

FIGURE 4.1
Classifications of relationships considered by ISA.

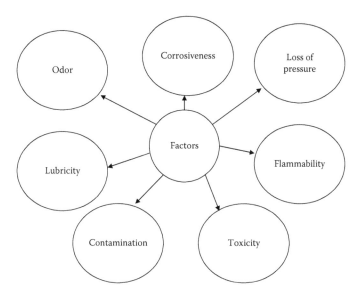

FIGURE 4.2
Factors that must be considered seriously from the safety aspect in the case of fluid.

The functional relationships are concerned with multiple items/units. For example, in a situation where an item's/unit's outputs constitute the inputs to a downstream item/unit, an error in inputs and outputs may cause damage to the downstream item/unit and, in turn, become a safety hazard. The condition of outputs could be excessive outputs, degraded outs, erratic outputs, unprogrammed outputs, or zero outputs.

Finally, the flow relationships may involve two or more items/units. For example, flow between two items/units may entail steam, air, lubricating oil, water, fuel, or electrical energy. Furthermore, the flow also could be unconfined, such as heat radiation from one item/body to another. Generally, the common problems associated with many products are the appropriate flow of energy and fluids from one unit/item to another through confined spaces/passages, consequently leading to direct or indirect safety-associated problems/issues. Nonetheless, the flow-related problem causes include faulty connections between units/items and complete or partial interconnection failure. In the case of fluid, factors (such as those shown in Figure 4.2) must be considered seriously from the safety aspect [14,17].

4.5 Task Analysis

This is a systematic approach and is used for assessing the needs of equipment maintainers for effectively working with hardware to conduct a stated task. The analyst, in addition to making careful observations regarding impediments to effective maintainability, documents and oversees each task element and start and completion times. The observations are categorized under the following 16 classifications [3,18]:

- **Classification 1:** Decision-making factors
- **Classification 2:** Training needs
- **Classification 3:** Equipment damage potential
- **Classification 4:** Maintenance crew interactions
- **Classification 5:** Lifting or movement aids
- **Classification 6:** Equipment maintainability design features
- **Classification 7:** Tools and job aids
- **Classification 8:** Personnel hazards
- **Classification 9:** Facility design features
- **Classification 10:** Spare parts retrieval
- **Classification 11:** Availability of necessary maintenance-related information (e.g., manuals, schematics, and procedures)
- **Classification 12:** Communication
- **Classification 13:** Environmental factors
- **Classification 14:** Supervisor-subordinate relationships
- **Classification 15:** Access factors
- **Classification 16:** Workshop adequacy

Additional information on this method is available in Reference 19.

4.6 Hazards and Operability Analysis (HAZOP)

This method was developed for application in the chemical industrial sector and is considered a quite powerful tool to highlight safety-related problems prior to availability of complete sets of data concerning an item in question [14,20]. Three fundamental objectives of this method are as follows [11,21,22]:

i. To produce a complete process/facility description
ii. To review each and every process/facility part for determining how deviations from the design intention can occur
iii. To decide whether such deviations can result in operating-related problems/hazards

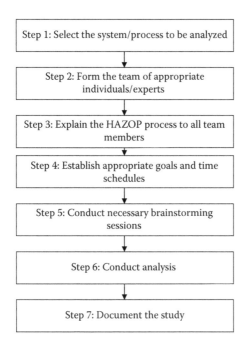

FIGURE 4.3
Steps involved in performing HAZOP.

The seven steps involved in performing hazards and operability analysis (HAZOP) are shown in Figure 4.3 [11,14,23].

Finally, it should be noted that this method (i.e., HAZOP) has basically the same weaknesses as FMEA discussed below in Section 4.8. For example, both HAZOP and FMEA predict problems that are connected to process/system-related failures but fail to factor human error into the equation. This is an important weakness because quite frequently in the occurrence of accidents, human error is a factor.

4.7 Pontecorvo Method

In nuclear power plants, this method can be used to obtain reliability estimates of task performance by a worker. The method initially obtains reliability estimates for discrete and separate subtasks with no accurate reliability values; it combines these very values to obtain the overall task reliability. Usually, this method is used during the initial phases of design for quantitatively assessing the interaction of machines and humans. The method can also be utilized for determining the performance of a single individual acting alone.

The Pontecorvo method is composed of the following six steps [4,24]:

- **Step 1: Identify tasks.** This step is concerned with highlighting the tasks to be performed. All these tasks should be highlighted at a gross level (i.e., one entire operation is to be represented by each task).
- **Step 2: Highlight subtasks of each task.** This step is concerned with the identification of each task's subtasks that are essential for its completion.
- **Step 3: Obtain empirical performance-related data.** This step is concerned with collecting empirical performance-related data from various sources including in-house operations and experimental literature. It should be noted that these data values should be subject to those types of environments under which subtasks are to be carried out.
- **Step 4: Establish subtask rate.** This step is concerned with rating each subtask as per its level of perceived difficulty or potential for the error occurrence. Generally, a 10-point scale is utilized to judge the appropriate subtask rate. The scale varies from most error to least error.
- **Step 5: Predict subtask reliability.** This step is concerned with predicting subtask reliability and is accomplished by expressing the judged ratings of the data as well as the empirical data in the form of a straight line. For goodness of fit, the regression line is tested.
- **Step 6: Determine task reliability.** This step is concerned with determining the task reliability, which is obtained by multiplying the reliabilities of subtasks.

It should be noted that the above-described approach is utilized to estimate the performance of a single individual acting alone. However, in a situation when a backup person is available, the probability of the task being performed properly (i.e., the task reliability) would be greater. In such a situation, the overall reliability of two individuals working together to carry out a stated task can be estimated by using the equation presented below [4,24]:

$$R_{ot} = \frac{\left[\left\{1-(1-R_s)^2\right\}P_{ba} + R_s P_{bu}\right]}{(P_{ba} + P_{bu})} \tag{4.1}$$

where

R_{ot} is the overall reliability of two individuals working together to perform a given task.

R_s is the reliability of the single person/individual.

P_{ba} is the percentage of time the backup person/individual is available.

P_{bu} is the percentage of time the backup person/individual is unavailable.

EXAMPLE 4.1

Assume that in two workers are working together in a nuclear power plant to perform an operation-related task. The backup worker is available about 90% of the time, and the reliability of each worker is 0.85.

Calculate the reliability of performing the operation-related task correctly.

By inserting the specified data values into Equation 4.1, we obtain

$$R_{ot} = \left[\frac{\{1 - (1 - 0.85)^2\}0.9 + (0.85)(0.1)}{(0.9 + 0.1)} \right] = 0.9647$$

Thus, the reliability of performing the operation-related task correctly is 0.9647.

4.8 Failure Modes and Effect Analysis (FMEA)

This is probably the most widely used method during the design process for analyzing engineering systems from their reliability aspect. FMEA may simply be described as an effective approach for analyzing each potential failure mode in the system to determine the effects of such failure modes on the entire system [25]. When the effect of each failure mode is classified according to its severity, FMEA is referred to as failure mode effects and criticality analysis (FMECA).

The history of FMEA may be traced back to the early 1950s with the development of flight control systems, when the Bureau of Aeronautics of the U.S. Navy, in order to develop a procedure for reliability control over the detail design effort, developed a requirement called Failure Analysis [26]. Subsequently, the term "Failure Analysis" was changed to FMEA, and in the 1970s, the U.S. Department of Defense directed its effort to developing a military standard entitled "Procedures for Performing a Failure Mode, Effects, and Criticality Analysis" [27].

Basically, FMECA is an extended version of FMEA. More clearly, when FMEA is extended to group each potential failure effect with respect to its severity (this includes documenting catastrophic and critical failures), the method is referred to as FMECA [14,28].

The seven main steps followed to perform FMEA are as follows [2,7,14]:

- **Step 1:** Define system boundaries and its associated requirements.
- **Step 2:** List system subsystems and components.
- **Step 3:** Identify and describe each component and list its failure modes.
- **Step 4:** Assign probabilities/failure rates to each component's failure modes.

- **Step 5:** List each failure mode's effect/effects on subsystems, system, and plant.
- **Step 6:** Enter remarks for all failure modes.
- **Step 7:** Review each critical failure mode and take appropriate action.

There are many factors that must be explored with care prior to the implementation of FMEA. Some of these factors are as follows [29,30]:

- Review all conceivable failure modes by the involved professionals.
- Making decisions based on the risk priority number (RPN).
- Obtaining engineer's clear approval and support.
- Measuring with care FMEA cost/benefits.

Over the years professionals working in the area of reliability analysis have developed a number of facts/guidelines concerning this method (i.e., FMEA). Some of these facts/guidelines are as follows [1,14,29]:

- FMEA has certain limitations.
- RPN could be directly or indirectly misleading.
- FMEA is not designed to supersede the work of an engineer.
- FMEA is not the method for choosing the optimum design concept.
- Avoid developing most of the FMEA in a meeting.

There are many advantages of performing FMEA. Some of these advantages are that it identifies safety concerns to be focused on, is easy to understand, is a systematic approach for classifying hardware failures, provides a safeguard against repeating the same mistakes in the future, is a useful approach that starts from the detailed level and works upward, is a visibility tool for management, improves customer satisfaction, is a useful approach for improving communication among design interface personnel, is a useful approach for comparing designs, serves as a useful tool for more efficient test planning, reduces engineering changes, and reduces development cost and time [2,14,29].

Additional information on this method is available in Reference 2.

4.9 Man-Machine Systems Analysis

This method was developed in the early years of the 1950s for reducing human error–caused unwanted effects to some acceptable level in a system. The method is composed of the following 10 steps [3,31]:

- **Step 1:** Define system goals and its functions.
- **Step 2:** Define concerned situational-related characteristics (i.e., the performance-shaping factors such as illumination and air quality under which tasks have to be carried out).
- **Step 3:** Define all involved individuals' characteristics (e.g., motivation, skills, experience, and training).
- **Step 4:** Define jobs performed by all involved individuals.
- **Step 5:** Analyze jobs with regard to identifying potential error-likely situations and other related difficulties.
- **Step 6:** Estimate chances or other information with regard to potential human error occurrence.
- **Step 7:** Determine the likelihood of failure for detecting and rectifying a potential human error.
- **Step 8:** Determine the possible consequences of failure for detecting potential human errors.
- **Step 9:** Recommend appropriate changes.
- **Step 10:** Reevaluate each change by repeating with care most of the above steps.

Additional information on this method is available in Reference 31.

4.10 Maintenance Personnel Performance Simulation (MAPPS) Model

This model's development was sponsored by the U.S. Nuclear Regulatory Commission (NRC), and it was developed by the Oak Ridge National Laboratory for providing estimates of performance measures of nuclear power plant maintenance man power [32]. The main objective for the model's development was the pressing need for and lack of a human reliability-related data bank pertaining to nuclear power plant maintenance-related tasks for application in carrying out probabilistic risk assessment-related studies.

There are many performance measures estimated by the maintenance personnel performance simulation (MAPPS) model. Four of these performance measures are as follows [3]:

- Probability of successfully performing the task of interest
- Identification of the least and most likely error-prone sub elements

- Maintenance team stress-related profiles during the task execution period
- Probability of an undetected error

Finally, it is added that MAPPS model is a very good tool for estimating important maintenance-related parameters, and its flexibility allows it to be employed in a quite wide range of studies concerning nuclear power plant maintenance activity.

Additional information on this model is available in Reference 32.

4.11 Fault Tree Analysis (FTA)

This is a widely used method for performing reliability-related analysis of engineering systems in the industrial sector, particularly in the area of nuclear power generation. The method was developed in the early 1960s for performing reliability-related analysis of the Minuteman Launch Control System [2,7].

A fault tree may simply be described as a logical representation of the relationship of primary/basic fault events that lead to the occurrence of a specified undesirable event called the "top event," and is depicted by using a tree structure with logic gates such as OR and AND. Although there could be many purposes in conducting FTA, the main ones include understanding the level of protection that the design concept provides against failures, identifying critical areas and cost-effective improvements, satisfying jurisdictional-related requirements, and understanding the functional relationship of system failures.

FTA starts by identifying an undesirable event, called the "top event," associated with a system under consideration. Fault events that can cause its occurrence (i.e., the top event) are connected and generated by logic operators, such as OR and AND. The construction of a fault tree proceeds by generating fault events in a successive manner until the fault events need not be developed any further. During the construction process of a fault tree, one successively asks the question: "How could this fault event occur?"

There are many symbols used in the construction of fault trees. The four basic ones are shown in Figure 4.4 and the information on other symbols is available in References 2, 7, and 33.

All symbols shown in Figure 4.4 are described below.

- **OR gate:** It denotes that an output fault event occurs if one or more of the input fault events occur.
- **AND gate:** It denotes that an output fault event occurs only if all of the input fault events occur.

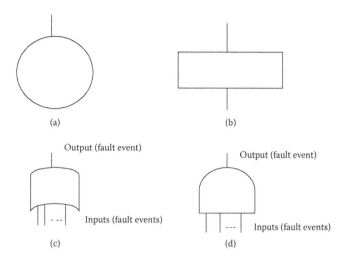

FIGURE 4.4
Basic fault tree symbols: (a) Circle, (b) Rectangle, (c) OR gate, (d) AND gate.

- **Rectangle:** It represents a fault event that occurs from the logical combination of fault events through the input of a logic gate, such as OR and AND.
- **Circle:** It represents a basic fault event (e.g., failure of an elementary component or part). The parameters of the event are probability of occurrence, failure and repair rates (the values of these parameters are normally obtained from empirical data).

Normally, the steps presented below are followed in performing FTA [2,3,34].

- **Step 1:** Define the system and analysis-related assumptions.
- **Step 2:** Identify the system undesirable fault event (i.e., the system top fault event to be investigated).
- **Step 3:** Highlight all the possible causes that can make the system top fault event to occur by utilizing fault tree symbols and the logic tree format.
- **Step 4:** Develop the fault tree to the lowest level of detail as per the stated requirements.
- **Step 5:** Conduct analysis of the completed fault tree with regard to factors such as gaining proper insight into the unique modes of product-related faults and comprehending the proper logic and the interrelationships among various fault paths.
- **Step 6:** Determine the most appropriate corrective measures.
- **Step 7:** Document the analysis and follow up on all the highlighted corrective measures.

EXAMPLE 4.2

Assume that a nuclear power plant operator is required to perform task T. The task is composed of two subtasks: M and N. If any one of these two subtasks is performed incorrectly, the task T will be performed incorrectly.

Subtask M is composed of two steps: *a* and *b*. If any one of these two steps is performed incorrectly, the subtask M will be performed incorrectly. Subtask N is composed of three steps: *x*, *y*, and *z*. All these three steps must be performed incorrectly for subtask N to be performed incorrectly.

If subtasks and steps are independent, develop a fault tree by using Figure 4.4 symbols for the undesired event (i.e., top event): the nuclear power plant operator will not perform the task, T, correctly.

By using Figure 4.4 symbols, the fault tree shown in Figure 4.5, for this example, is developed. Each fault event is labeled as E_1, E_2, E_3, E_4, E_5, E_6, E_7 and E_8.

4.11.1 Fault Tree Probability Evaluation

Under certain scenarios, it may be necessary to predict the probability of occurrence of a certain top event (e.g., the nuclear power plant operator will not perform the task, T, correctly). Before this could be achieved by utilizing the FTA method, the determination of the probability of occurrence of output fault events of all involved logic gates is required.

Thus, the occurrence probability of the output fault event of an OR gate is expressed by [2,7]:

$$P_o(y) = 1 - \prod_{i=1}^{m}(1 - P(y_i)) \tag{4.2}$$

where

$P_o(y)$ is the probability of occurrence of OR gate's output fault event *y*.
m is the number of input fault events.
$P(y_i)$ is the probability of occurrence of input fault event y_i, for $i = 1, 2, 3, \ldots, m$.

Similarly, the occurrence probability of the output fault event of an AND gate is expressed by

$$P_a(y) = \prod_{i=1}^{m} P(y_i) \tag{4.3}$$

where

$P_a(y)$ is the probability of occurrence of AND gate's output fault event *y*.

EXAMPLE 4.3

Assume that the probabilities of occurrence of fault events E_4, E_5, E_6, E_7, and E_8 in Figure 4.5 fault tree are 0.01, 0.02, 0.03, 0.04, and 0.05, respectively.

Calculate the probability of occurrence of the top fault event (i.e., the nuclear power plant operator will not perform the task, T, correctly), and then redraw the Figure 4.5 fault tree with the calculated and given fault event occurrence probability values.

By substituting the given data values into Equation 4.2, we obtain the following probability value for the occurrence of fault event E_2 (i.e., the subtask, M, will not be performed correctly):

$$P(E_2) = 1 - \{1 - P(E_4)\}\{1 - P(E_5)\}$$
$$= 1 - \{1 - 0.01\}\{1 - 0.02\}$$
$$= 0.0298$$

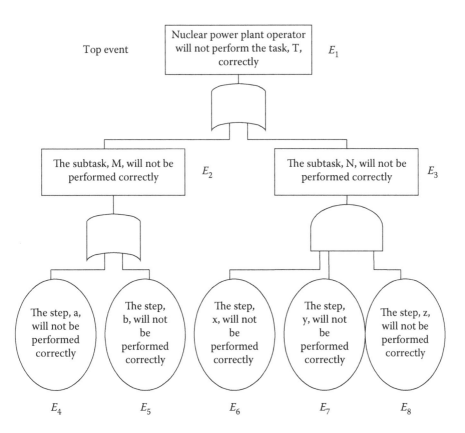

FIGURE 4.5
A fault tree for the unsuccessful performance of task, T, by the nuclear power plant operator.

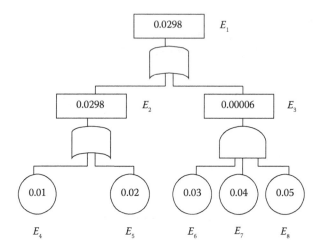

FIGURE 4.6
Redrawn Figure 4.5 fault tree with the calculated and given fault event occurrence probability values.

Similarly, by substituting the given data values into Equation 4.3, we obtain the following probability value for the occurrence of fault event E_3 (i.e., the subtask, N, will not be performed correctly):

$$P(E_3) = P(E_6)P(E_7)P(E_8)$$
$$= (0.03)(0.04)(0.05)$$
$$= 0.00006$$

By substituting the above two calculated probability values into Equation 4.2, we obtain the following value for the occurrence of the top event, E_1 (i.e., the nuclear power plant operator will not perform the task, T, correctly):

$$P(E_1) = 1 - \{1 - P(E_2)\}\{1 - P(E_3)\}$$
$$= 1 - \{1 - 0.0298\}\{1 - 0.00006\}$$
$$= 0.0298$$

Thus, the probability of occurrence of the top fault event (i.e., the nuclear power plant operator will not perform the task, T, correctly) is 0.0298. The fault tree with the calculated and given fault event occurrence probability values is shown in Figure 4.6.

4.11.2 FTA Benefits and Drawbacks

Just like any other reliability analysis method, the FTA method too has its benefits and drawbacks. Thus, some of its benefits are as follows [2,7]:

- Useful for providing insight into the system behavior and for identifying failures deductively.
- Useful for handling complex systems more easily.
- A graphic aid for management personnel.
- Useful for providing options for management and others to perform either quantitative or qualitative analysis.
- Useful because it requires the involved analyst to comprehend thoroughly the system under consideration before starting the analysis.
- Useful because it allows concentration on one specific failure at a time.

In contrast, some drawbacks of the FTA method are as follows [2,7]:

- Its final results are difficult to check.
- It considers parts in either working or failed state. More specifically, the parts' partial failure states are difficult to handle.
- It is a costly and a time-consuming method.

Additional information on this method is available in References 2 and 7.

4.12 Markov Method

This is a widely used method for performing various types of reliability analysis of engineering systems, and it was developed by a Russian mathematician named Andrei A. Markov (1856–1922). Thus, it is a very useful tool for performing various types of analysis concerning safety, reliability, and human error in nuclear power plants.

The following three assumptions are associated with the Markov method [2,35]:

- The transitional probability from one system state to another in finite time interval Δt is given by $\lambda \Delta t$, where λ is the transition rate (e.g., failure rate) from one system state to another.
- The probability of more than one transition occurrence in the finite time interval Δt from one system state to another is negligible (e.g., $(\lambda \Delta t)(\lambda \Delta t) \rightarrow 0$).
- All occurrences are independent of each other.

FIGURE 4.7
Nuclear power plant system state space diagram.

The application of the Markov method to perform reliability analysis in nuclear power plants is demonstrated through the following example:

EXAMPLE 4.4

Assume that the constant failure rate of a system used in a nuclear power plant, is λ_s. The system state space diagram is shown in Figure 4.7. The numerals in boxes denote system states. Develop expressions for the system state probabilities and mean time to failure by using the Markov method and assuming that the system failures occur independently.

By using the Markov method, we write the following equations for the state space diagram shown in Figure 4.7 [2,35]:

$$P_0(t + \Delta t) = P_0(t)(1 - \lambda_s \Delta t) \tag{4.4}$$

$$P_1(t + \Delta t) = P_0(t)(\lambda_s \Delta t) + P_1(t)(1 - 0\Delta t) \tag{4.5}$$

where
λ_s is the constant failure rate of the nuclear power plant system.
$\lambda_s \Delta t$ is the probability of the nuclear power plant system failure in finite time interval Δt.
$P_0(t + \Delta t)$ is the probability of the nuclear power plant system being in operating state 0 at time $(t + \Delta t)$.
$P_1(t + \Delta t)$ is the probability of the nuclear power plant system being in failed state 1 at time $(t + \Delta t)$.
$(1 - \lambda_s \Delta t)$ is the probability of no nuclear power plant system failure in finite time interval Δt.
$P_i(t)$ is the probability that the nuclear power plant system is in state i at time t, for $i = 0$ (operating normally) and $i = 1$ (failed).

By rearranging Equations 4.4 and 4.5 and taking the limit as $\Delta t \to 0$, we get

$$\lim_{\Delta t \to 0} \frac{P_0(t + \Delta t) - P_0(t)}{\Delta t} = \frac{dP_0(t)}{dt} = -\lambda_s P_0(t) \tag{4.6}$$

$$\lim_{\Delta t \to 0} \frac{P_1(t + \Delta t) - P_1(t)}{\Delta t} = \frac{dP_1(t)}{dt} = \lambda_s P_0(t) \tag{4.7}$$

At time $t = 0$, $P_0(0) = 1$ and $P_1(0) = 0$.

By solving Equations 4.6 and 4.7, with the aid of Laplace transforms, we get

$$P_0(s) = \frac{1}{s + \lambda_s} \tag{4.8}$$

$$P_1(s) = \frac{\lambda_s}{s(s + \lambda_s)} \tag{4.9}$$

where
s is the Laplace transform variable.

Taking the inverse Laplace transforms of Equations 4.8 and 4.9, we obtain

$$P_0(t) = e^{-\lambda_s t} \tag{4.10}$$

$$P_1(t) = 1 - e^{-\lambda_s t} \tag{4.11}$$

Equations 4.10 and 4.11 are the expressions for the system state probabilities. Equation 4.10 is the reliability of the nuclear power plant system. In other words

$$R_{nps}(t) = P_0(t) = e^{-\lambda_s t} \tag{4.12}$$

where
$R_{nps}(t)$ is the reliability of the nuclear power plant system at time t.

By integrating Equation 4.12 over the time interval $[0, \infty]$, we get the following expression for the nuclear power plant system mean time to failure [2]:

$$\begin{aligned} MTTF_{nps} &= \int_0^\infty R_{nps}(t)dt \\ &= \int_0^\infty e^{-\lambda_s t}dt \\ &= \frac{1}{\lambda_s} \end{aligned} \tag{4.13}$$

where
$MTTF_{nps}$ is the mean time to failure of the nuclear power plant system.

EXAMPLE 4.5

Assume that the constant failure rate of a system used in a nuclear power plant is 0.0002 failures per hour. Calculate the system mean time to failure and reliability during an 800-hour mission.

By substituting the specified data values into Equations 4.13 and 4.12, we get

$$MTTF_{nps} = \frac{1}{0.0002} = 5000 \text{ hours}$$

and

$$R_{nps}(800) = e^{-(0.0002)(800)} = 0.8521$$

Thus, the system mean time to failure and reliability are 5000 hours and 0.8521, respectively.

4.13 Probability Tree Method

This method is used for performing task analysis by diagrammatically representing critical human-related actions and other events associated with the system under consideration. Often, the method is used for performing task analysis in the technique for human error rate prediction (THERP) [4,36].

In this method, diagrammatic task analysis is denoted by the branches of the probability tree. The branching limbs of the tree denote each event's outcome (i.e., success or failure) and each branch is assigned probability of occurrence.

There are many benefits of this method including simplified mathematical computations, an effective visibility tool, and a very good tool for predicting the quantitative effects of errors [2,36].

The application of this method to a nuclear power plant system-related problem is demonstrated through the following example:

EXAMPLE 4.6

Assume that a nuclear power plant operator performs three independent tasks: x, y, and z. Task x is performed before task y, and task y before task z, and each of these three tasks can be performed either correctly or incorrectly.

Develop a probability tree for this example and obtain an expression for the probability of failure to accomplish the overall mission by the nuclear power plant operator.

In this case, the operator first performs task x correctly or incorrectly and then proceeds to perform task y. Task y can also be performed correctly or incorrectly by the operator. After task y, the operator proceeds to perform task z. This task can also be performed correctly or incorrectly

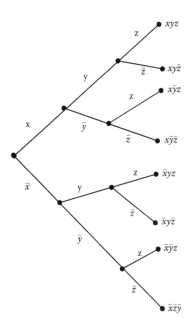

FIGURE 4.8
Probability tree diagram for the nuclear power plant operator performing tasks x, y, and z.

by the operator. This entire scenario is depicted by the probability tree diagram shown in Figure 4.8.

The six symbols used in Figure 4.8 are defined below.

x denotes the event that task x is performed correctly.
$\bar{x}$ denotes the event that task x is performed incorrectly.
y denotes the event that task y is performed correctly.
$\bar{y}$ denotes the event that task y is performed incorrectly.
z denotes the event that task z is performed correctly.
$\bar{z}$ denotes the event that task z is performed incorrectly.

By examining Figure 4.8, it can be concluded that there are seven distinct possibilities (i.e., $xy\bar{z}, x\bar{y}z, x\bar{y}\bar{z}, \bar{x}yz, \bar{x}y\bar{z}, \bar{x}\bar{y}z, \bar{x}\bar{y}\bar{z}$) for not successfully accomplishing the overall mission by the nuclear power plant operator. Thus, the probability of failure to accomplish the overall mission by the nuclear power plant operator is given by

$$P_{fo} = P(xy\bar{z}) + P(x\bar{y}z) + P(x\bar{y}\bar{z}) + P(\bar{x}yz) + P(\bar{x}y\bar{z}) + P(\bar{x}\bar{y}z) + P(\bar{x}\bar{y}\bar{z})$$
$$= P_x P_y P_{\bar{z}} + P_x P_{\bar{y}} P_z + P_x P_{\bar{y}} P_{\bar{z}} + P_{\bar{x}} P_y P_z + P_{\bar{x}} P_y P_{\bar{z}} + P_{\bar{x}} P_{\bar{y}} P_z + P_{\bar{x}} P_{\bar{y}} P_{\bar{z}} \quad (4.14)$$

where
P_{fo} is the probability of failure to accomplish the overall mission by the nuclear power plant operator.

P_x is the probability of performing task x correctly by the nuclear power plant operator.

$P_{\bar{x}}$ is the probability of performing task x incorrectly by the nuclear power plant operator.

P_y is the probability of performing task y correctly by the nuclear power plant operator.

$P_{\bar{y}}$ is the probability of performing task y incorrectly by the nuclear power plant operator.

P_z is the probability of performing task z correctly by the nuclear power plant operator.

$P_{\bar{z}}$ is the probability of performing task z incorrectly by the nuclear power plant operator.

Thus, Equation 4.14 is the expression for the failure to accomplish the overall mission by the nuclear power plant operator.

By substituting $P_x = 1 - P_{\bar{x}}, P_y = 1 - P_{\bar{y}}$, and, $P_z = 1 - P_{\bar{z}}$ into Equation 4.14, we get

$$P_{fo} = (1 - P_{\bar{x}})(1 - P_{\bar{y}})P_{\bar{z}} + (1 - P_{\bar{x}})P_{\bar{y}}(1 - P_{\bar{z}}) + (1 - P_{\bar{x}})P_{\bar{y}}P_{\bar{z}}$$
$$+ P_{\bar{x}}(1 - P_{\bar{y}})(1 - P_{\bar{z}}) + P_{\bar{x}}(1 - P_{\bar{y}})P_{\bar{z}} + P_{\bar{x}}P_{\bar{y}}(1 - P_{\bar{z}}) + P_{\bar{x}}P_{\bar{y}}P_{\bar{z}} \qquad (4.15)$$

EXAMPLE 4.7

Assume that in Example 4.5, the probabilities of failure to accomplish tasks x, y, and z by the nuclear power plant operator are 0.08, 0.04, and 0.06, respectively. Calculate the probability of failure to accomplish the overall mission by the nuclear power plant operator.

By substituting the given data values into Equation 4.15, we get

$$P_{fo} = (1 - 0.08)(1 - 0.04)(0.06) + (1 - 0.08)(0.04)(1 - 0.06)$$
$$+ (1 - 0.08)(0.04)(0.06) + (0.08)(1 - 0.04)(1 - 0.06) + (.08)(1 - 0.04)(0.06)$$
$$+ (0.08)(0.04)(1 - 0.06) + (0.08)(0.04)(0.06)$$
$$= 0.0529 + 0.0345 + 0.0022 + 0.0721 + 0.0046 + 0.0030 + 0.0001$$
$$= 0.1694$$

Thus, the probability of failure to accomplish the overall mission by the nuclear power plant operator is 0.1694.

4.14 Problems

1. Describe TOR.
2. What are the benefits and drawbacks of RCA?

3. Describe ISA.
4. Compare task analysis with HAZOP.
5. Describe the Pontecorvo method.
6. What is the difference between FMEA and FMECA?
7. What are the advantages of FMEA?
8. Describe the following two methods:
   a. Man-machine systems analysis
   b. MAPPS model
9. What are the benefits and drawbacks of performing FTA?
10. Prove Equations 4.8 and 4.9 by using Equations 4.6 and 4.7 and then prove Equations 4.10 and 4.11 by using Equations 4.8 and 4.9. Also find the sum of Equations 4.8 and 4.9.

References

1. Dhillon, B.S., *Engineering Safety: Fundamentals, Techniques, and Applications*, World Scientific Publishing, River Edge, New Jersey, 2003.
2. Dhillon, B.S., *Design Reliability: Fundamentals and Applications*, CRC Press, Boca Raton, Florida, 1999.
3. Dhillon, B.S., *Human Reliability, Error, and Human Factors in Power Generation*, Springer, London, 2014.
4. Dhillon, B.S., *Human Reliability: With Human Factors*, Pergamon Press, New York, 1986.
5. Dhillon, B.S., *Human Reliability, Error, and Human Factors in Engineering Maintenance: With Reference to Aviation and Power Generation*, CRC Press, Boca Raton, Florida, 2009.
6. Fault Tree Handbook, Report No. NUREG-0492, United States Nuclear Regulatory Commission, Washington, DC, 1981.
7. Dhillon, B.S., Singh, C., *Engineering Reliability: New Techniques and Applications*, John Wiley & Sons, New York, 1981.
8. Dhillon, B.S., *Patient Safety: An Engineering Approach*, CRC Press, Boca Raton, Florida, 2012.
9. Dhillon, B.S., *Engineering and Technology Management Tools and Applications*, Artechhouse, Inc., Norwood, Massachusetts, 2002.
10. Hallock, R.G., Technique of operations review analysis: Determine cause of accident/incident, *Safety and Health*, Vol. 60, No. 8, 1991, pp. 38–39.
11. Goetsch, D.L., *Occupational Safety and Health*, Prentice Hall, Englewood Cliffs, New Jersey, 1996.
12. Latino, R.J., Automating root cause analysis, in *Error Reduction in Health Care*, edited by P.L. Spath, John Wiley & Sons, New York, 2000, pp. 155–164.
13. Busse, D.K., Wright, D.J., Classification and Analysis of Incidents in Complex Medical Environments, Report. Available from the Intensive Care Unit, Western General Hospital, Edinburgh, UK, 2000.

14. Dhillon, B.S., *Safety and Human Error in Engineering Systems*, CRC Press, Boca Raton, Florida, 2013.
15. Burke, A., *Root Cause Analysis*, Report. Available from the Wild Iris Medical Education, P.O. Box 257, Comptche, California, 2000.
16. Wald, H., Shojania, K.G., Root cause analysis, in *Making Health Care Safer: A Critical Analysis of Patient Safety Practices*, edited by A.J., Markowitz, Report No. 43, Agency for Health Care Research and Quality, U.S. Department of Health and Human Services, Rockville, Maryland, 2001, Chapter 4, pp. 1–7.
17. Hammer, W., *Product Safety Management and Engineering*, Prentice Hall, Englewood Cliffs, New Jersey, 1980.
18. Seminara, J.L., Parsons, S.O., Human factors engineering and power plant maintenance, *Maintenance Management International*, Vol. 6, 1985, pp. 33–71.
19. Seminara, J.L., Human Factors Methods for Assessing and Enhancing Power Plant Maintainability, Report No. EPRI NP-2360, Electric Power Research Institute, Palo Alto, California, 1982.
20. *Guidelines for Hazard Evaluation Procedures*, American Institute of Chemical Engineers, New York, 1985.
21. Gloss, D.S., Wardle, M.G., *Introduction to Safety Engineering*, John Wiley & Sons, New York, 1984.
22. Roland, H.E., Moriarity, B., *System Safety Engineering and Management*, John Wiley & Sons, New York, 1983.
23. Risk Analysis Requirements and Guidelines, Report No. CAN/CSA-Q634-91, prepared by the Canadian Standards Association. Available from Canadian Standards Association, 178 Rexdale Blvd., Rexdale, Ontario, Canada, 1991.
24. Pontecorvo, A.B., A method of predicting human reliability, *Proceedings of the 4th Annual Reliability and Maintainability Conference*, Washington, DC, 1965, pp. 337–342.
25. Omdahl, T.P., ed., *Reliability, Availability, and Maintainability (RAM) Dictionary*, American Society for Quality Control (ASQC) Press, Milwaukee, Wisconsin, 1988.
26. MIL-F-18372 (Aer), *General Specification for Design, Installation, and Test of Aircraft Flight Control Systems*, Bureau of Naval Weapons, U.S. Department of the Navy, Washington, DC, 1987.
27. MIL-STD-1629, *Procedures for Performing a Failure Mode, Effects, and Criticality Analysis*, U.S. Department of Defense, Washington, DC, 1980.
28. Jordan, W.E, Failure modes, effects, and criticality analyses, *Proceedings of the Annual Reliability and Maintainability Symposium*, 1972, pp. 30–37.
29. Palady, P., *Failure Modes and Effects Analysis*, PT Publications, West Palm Beach, Florida, 1995.
30. McDermott, R.E., Mikulak, R.J., Beauregard, M.R., *The Basic FMEA*, Quality Resources, New York, 1996.
31. Miller, R.B., A Method for Man-Machine Task Analysis, Report No. 53-137, Wright Air Development Center, Wright-Patterson Air Force Base, U.S. Air Force (USAF), Ohio, 1953.
32. Knee, H.E., The Maintenance Personnel Performance Simulation (MAPPS) model: A human reliability analysis tool, *Proceedings of the International Conference on Nuclear Power Plant Aging, Availability Factors, and Power Plants*, 1988, pp. 373–377.

33. Schroder, R.J., Fault tree for reliability analysis, *Proceedings of the Annual Symposium on Reliability*, 1970, pp. 179–174.
34. Grant Ireson, W., Coombs, C.F., Moss, R.Y., eds., *Handbook of Reliability Engineering and Management*, McGraw-Hill, New York, 1996.
35. Shooman, M.L., *Probabilistic Reliability: An Engineering Approach*, McGraw-Hill, New York, 1968.
36. Swain, A.D., A Method for Performing a Human Factors Reliability Analysis, Report No. SCR-685, Sandia Corporation, Albuquerque, New Mexico, August 1963.

5

Human Reliability Analysis Methods for Nuclear Power Stations

5.1 Introduction

Nowadays, nuclear power stations have become an important element in power generation around the globe as they generate approximately 16% of the world's electricity [1]. Humans play a pivotal role in nuclear power generation, and their reliability has become an important issue as human error can lead to nuclear power station accidents such as Three Mile Island and Chernobyl. Furthermore, a study of Licensee Events Reports (LERs) carried out by the U.S. Nuclear Regulatory Commission (NRC) clearly indicates that upward of around 65% of nuclear system failures directly or indirectly involve error to a certain degree [2,3].

Over the years for performing human reliability analysis (HRA) in nuclear power stations, many methods have been developed [4–7]. These methods include a technique for human event analysis (ATHEANA), technique for human error rate prediction (THERP), cognitive reliability and error analysis method (CREAM), and accident sequence evaluation program (ASEP). This chapter presents important aspects of human reliability methods and methods considered useful for performing HRA in nuclear power stations.

5.2 Incorporation of the HRA Integrally into a Probabilistic Risk Assessment (PRA) and Human Reliability Method Requirements

HRA is an integral part of probabilistic risk assessment (PRA) as the risk associated with a nuclear power plant/station is very much dependent on the involved individuals' reliability and their interactions with the equipment and controls in the plant/station. Thus, the incorporation of HRA integrally into a PRA includes the following four issues [8]:

- **Issue 1: The relationship of approach to results.** This issue is basically concerned with the way in which HRA is conducted, its philosophy, and the end results or the insights that may be obtained.

- **Issue 2: Flexibility.** This issue is concerned with the flexibility required in using behavioral science-related technologies associated with HRA. It should be noted that HRA is a relatively immature method; however, it is based on the scientific disciplines developed in behavioral-related sciences with their inherent uncertainties to a degree. All in all, HRA should be flexible to accommodate new findings and model developments effectively while structured enough to be tractable and repeatable properly.

- **Issue 3: PRA Compatibility.** This issue is concerned with the compatibility of an HRA with the PRA of which it is an element/component. Furthermore, it should be noted that the risk applications of PRA require the HRA process to specify clearly human interaction–associated events in qualitative detail, adequate for guiding effectively the risk management-related efforts from the human factors perspective.

- **Issue 4: Limits.** This issue is concerned with the HRA limits or the limits of HRA end results. Thus, the end results of HRA should be documented in such a way that is easily understandable to general PRA users. Furthermore, the end results should be easily traceable to their models, basic assumptions, and data sources.

As HRA forms a very important component of nuclear power stations' probabilistic safety assessment-related studies, the probabilistic safety assessment objective ought to be clearly reflected in the choice of the HRA method [9]. This way, if the probabilistic safety assessment were to be employed by a regulatory body personnel for assessing the general risk of the industry, one would make use of generic methods/techniques representative of a nominal nuclear power station. Needless to say, the quality of HRA is one of the requirements to setting probabilistic safety assessment application.

Furthermore, quality is reflected in both what method/technique is chosen and how it is employed. For achieving a "good" quality rating, one has to tie the method/technique to the application of a nongeneric method/technique supported by the use of power plant personnel (i.e., experts in their specializations) in addition to feedback received from plant experience and the application of the simulator. Additional information on requirements for an HRA method/technique is available in Reference 9.

5.3 HRA Process Steps and Their End Results

HRA process for use in nuclear power stations is composed of eight steps. These steps (along with their end results in parentheses) are as follows [8]:

- **Step 1: Select and train team personnel.** (Set of various types of skills embodied in an integrated team.)
- **Step 2: Familiarize all team personnel with nuclear power plant.** (Initial identification of human-related functions and activities. In addition, problems can be spotted, although without much risk context yet.)
- **Step 3: Develop initial nuclear power plant model.** (Most important human-related interactions highlighted, system interactions highlighted, all major systems modelled, and a measure of the defense barriers against major classes of off-normal events.)
- **Step 4: Screen human-related interactions.** (Important human-related interactions highlighted, initial quantification conducted, and screening values selected.)
- **Step 5: Characterize human-related interactions.** (Failure modes, causes, effects, mechanisms, and influences of the important human-related interactions are determined.)
- **Step 6: Quantify human-related interactions.** (Importance ranking, likelihood, and uncertainties of important human-related interactions.)
- **Step 7: Update nuclear power plant model.** (Interactions and recovery models added to model.)
- **Step 8: Review final results.** (Confidence that final/end results make sense and can be utilized by nuclear power plant personnel in risk applications.)

Additional information on HRA process steps and their final/end results is available in Reference 8.

5.4 HRA Methods

Over the years, many methods have been developed to perform various types of HRA [4,10]. Seven of these methods that are considered quite useful for performing HRA in nuclear power plants/stations are as follows [4,5,7,11–15]:

- Success likelihood index method–Multiattribute utility decomposition (SLIM-MAUD)

- THERP
- ASEP
- ATHEANA
- Human Error Assessment and Reduction Technique (HEART)
- CREAM
- Standardized plant analysis risk–Human reliability analysis (SPAR-HRA)

Thirteen information requirements of the above first three methods (i.e., SLIM-MAUD, THERP, and ASEP) are as follows [11,16]:

- **Information requirement 1:** The available procedures
- **Information requirement 2:** Appropriate description of the task and the action
- **Information requirement 3:** The time required to carry out the task appropriately
- **Information requirement 4:** The dependence of these different times
- **Information requirement 5:** The dependence of the different tasks
- **Information requirement 6:** The individuals or teams that have to carry out the task
- **Information requirement 7:** The available total time for diagnosis and execution of a task correctly (i.e., time window for action)
- **Information requirement 8:** Influence factors (e.g., fatigue, skill, and stress)
- **Information requirement 9:** Demands of cognition and action, the level of experience, and perception
- **Information requirement 10:** The error type
- **Information requirement 11:** The technical systems and their associated dynamics
- **Information requirement 12:** The working media or the man-machine interface
- **Information requirement 13:** Recovery factors for all the different tasks

Additional information on the above 13 requirements is available in Reference 16.

All of the above seven HRA methods are described below, separately.

5.4.1 Success Likelihood Index Method–Multiattribute Utility Decomposition (SLIM-MAUD)

For the U.S. NRC, this method was developed by British researchers as a means of automating some of the mechanics of the existing success likelihood index method [5,15,17]. SLIM-MAUD makes use of paired-comparison methods under the assumptions that similar tasks are categorized for analysis and upper and lower bound anchor point human error probabilities are determined for those categories of tasks. Performance-shaping factors are rated with regard to their importance and quality (and analysts' expert opinions are utilized for selecting the bounding human error probabilities) for determining the tasks to be similar enough to be compared as well to assess the performance-shaping factors.

Nonetheless, the SLIM-MAUD methodology breaks down into seven steps shown in Figure 5.1 [4,18].

The strength of this method lies in the paired-comparison procedure that enhances the analysis reliability. Furthermore, SLIM-MAUD comes

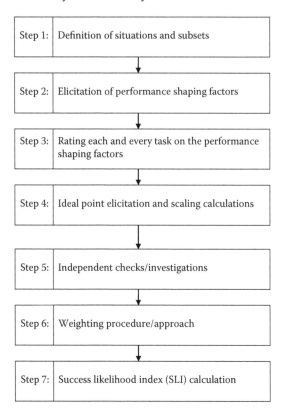

Step 1:	Definition of situations and subsets
Step 2:	Elicitation of performance shaping factors
Step 3:	Rating each and every task on the performance shaping factors
Step 4:	Ideal point elicitation and scaling calculations
Step 5:	Independent checks/investigations
Step 6:	Weighting procedure/approach
Step 7:	Success likelihood index (SLI) calculation

FIGURE 5.1
SLIM-MAUD methodology steps.

with a set of suggested performance-shaping factors, although all involved analysts are free to come up with their own. Nonetheless, SLIM-MAUD is considered atomistic due to its reliance on performance-shaping factors [5,15].

Additional information on this method is available in Reference 4.

5.4.2 Technique for Human Error Rate Prediction (THERP)

This was the first method in human reliability assessment to come into broad use. It was developed at the Sandia Laboratories for the U.S. NRC [19]. It is used in the area of human reliability assessment for evaluating the probability of a human error occurring throughout the completion of a specific task. The method is based upon both expert judgements and plant data; thus, it heavily relies on a large human reliability database that contains human error occurrence probabilities.

THERP is a total methodology for assessing human reliability that deals with task analyses (e.g., talk/walk through and documentation reviews), error identification and representation, and human error probabilities' quantification [20]. The key factors for completing the quantification process are decomposition of tasks into elements, assignment of nominal human error occurrence probabilities to all elements, determination of a performance-shaping factor's effects on each element, calculation of effects of dependence between tasks, modeling of a human reliability assessment event tree, and quantification of total task human error probability [4,21].

For obtaining the overall failure probability, the exact equation involves adding probabilities of all failure paths in the event tree under consideration.

As per References 4, 11, and 20, the main benefits and drawbacks of THERP are as follows:

Benefits
- It is well used in practice.
- It can be utilized at all stages of design.
- It has a powerful methodology that can be audited quite easily.

Drawbacks
- It fails to offer sufficient guidance on modeling scenarios as well as the performance-shaping factors' impact on performance.
- THERP can be quite time consuming as well as resource intensive.
- The level of detail included in THERP may be quite excessive for many assessments.

Additional information on THERP is available in Reference 4.

5.4.3 Accident Sequence Evaluation Program (ASEP)

This method was developed in 1987. It was designed as a simplified version of the THERP [4,22]. The unique features of ASEP include the following items [15,22]:

- Use of software and tables for providing uncertainty bounds.
- A quite detailed screening procedure for pre- and postaccident tasks.
- Consideration of the role of diagnosis in error and recovery.
- A fairly simplified three-level account of dependency.
- Tables for accounting for the influence of available time on the error probability.
- Inclusion of recovery-related factors.
- Separate human error probabilities for pre- and postaccident tasks.

The human error probability's handling characterizes the distinction between this method and THERP. More clearly, THERP requires the analyst to calculate the human error probability, whereas ASEP provides predefined human error probability values. This in turn decreases the analysis accuracy to the advantage of the simplicity and ease of carrying out the analysis. It should be noted that beyond open-ended performance-shaping factors, this method (i.e., ASEP) explicitly accounts for time, procedures, stress, and immediate response.

Additional points to be noted with regard to ASEP are as follows [5,15]:

- Its format is not a worksheet or rubric but a checklist procedure.
- The application of lookup tables for conducting calculations eradicates any ambiguity in probability assignments to events or in the overall calculation of error probability.
- It can be counted as an atomistic approach because of the degree/level of proceduralized detail given in it.

All in all, ASEP is a nuclear-specific method, and it has been quite successfully applied in the nuclear power industrial sector and additional information on ASEP is available in References 4, 5, 15, and 22.

5.4.4 A Technique for Human Event Analysis (ATHEANA)

This is a postincident HRA method. It was developed by the U.S. NRC for the following three main reasons [4,23,24]:

- Human events modeled in existing human reliability assessment/PRA models were considered to be quite inconsistent with the degree of roles that human operators have played in operational-related events.

- In comprehending human error, the accident record and advances made in behavioral sciences both strongly supported a stronger focus on the contextual-related factors, particularly plant conditions.
- Advances made in the area of psychology were fully integrated with the disciplines of engineering, human factors, and PRA when modeling human failure–associated events.

Seven fundamental steps to the ATHEANA methodology are presented below [7,24,25].

- **Step 1:** Define and interpret carefully the issue under consideration.
- **Step 2:** Define and detail the required analysis scope.
- **Step 3:** Describe carefully the base-case scenario including the norm of operations within the environments, considering actions as well as procedures.
- **Step 4:** Define human failure events and/or unsafe actions which may directly or indirectly affect the task under consideration.
- **Step 5:** Classify the highlighted human failure events under two basic groups: unsafe and safe actions. An unsafe action may be expressed as an action in which the human operator in question may fail to carry out a stated task or performs it incorrectly that in turn leads to the unsafe operation of the system.
- **Step 6:** Search carefully for any deviations from the base-case scenario with regard to any divergence in the normal environmental operating behavior in the context of the situational scenario.
- **Step 7:** Preparation for employing ATHEANA.

The probability of a human failure event occurring in ATHEANA is given by [7,12]:

$$P_{hf} = P_{ua}P_{nr}P_{ef} \tag{5.1}$$

where
P_{hf} is the probability of human failure event occurrence.
P_{ua} is the probability of unsafe action in the error-forcing context.
P_{nr} is the probability of nonrecovery in the error-forcing context and given the occurrence of the unsafe action as well as the existence of additional evidence after the unsafe action.
P_{ef} is the error-forcing context probability.

The main benefits and drawbacks of ATHEANA are as follows:

Benefits

- Its methodology makes it possible to estimate human error occurrence probabilities considering a variety of differing factors and combinations.
- It improves the chances that key risks concerning the human failure events in question have been highlighted.
- In comparison with many other HRA quantification methods, ATHEANA clearly allows for the consideration of a quite wider range of performance-related shaping factors, also it does not require that these factors to be treated as independent.
- In comparison with many other HRA methods, ATHEANA provides a much better as well as more holistic comprehension of the context concerning the human factors considered to be the cause of the incident.

Drawbacks

- ATHEANA fails to prioritize or establish appropriate details of the causal relationships between the types of human factors contributing to the occurrence of an incident.
- Another drawback of ATHEANA is that, from a probability risk assessment aspect, there is no human error probability generated.

Additional information on ATHEANA is available in Reference 4.

5.4.5 Human Error Assessment and Reduction Technique (HEART)

HEART is used in the area of human reliability assessment for evaluating the probability of human error occurrence throughout the completion of a certain task. It first appeared in 1986 and is widely used in the nuclear industrial sector of the United Kingdom [4,7,21,26,27]. HEART is based on the following three premises [4]:

- Basic human reliability very much depends on the generic nature of the task to be conducted.
- Under "perfect" conditions, this level of reliability will tend to be achieved quite consistently with a stated nominal likelihood within probabilistic limits.
- Given that the perfect conditions under all circumstances do not exist, the predicted human reliability may degrade as a function of the extent to which highlighted error generating conditions might be applicable.

In HEART, nine generic task types are described, each with an associated nominal human error occurrence potential and a total of 38 error-generating situations that may directly or indirectly affect reliability of task, each with a maximum amount by which the nominal human error occurrence potential can be multiplied. The key elements of HEART are as follows [4,7]:

- Categorize the task for analysis into any one of the nine generic task types and then assign the human error occurrence potential to that specific task.
- Make a decision about which error-producing conditions may directly or indirectly affect the reliability of the task, and then consider/evaluate the already assessed proportion of affect for all error-producing conditions.
- Calculate the task human error occurrence potential.

The main advantages and disadvantages of HEART are as follows [4,7,21,26,27]:

Advantages

- It is a quick, versatile, and simple and straightforward human reliability calculation method, which also provides the user useful suggestions for reducing human error.
- It needs a relatively limited amount of resources for completing an assessment.

Disadvantages

- It lacks information regarding the extent to which tasks should be decomposed for performing analysis.
- It does not incorporate error dependency modeling.
- It needs greater description clarity to assist potential users when discriminating between generic tasks and their related error-generating conditions.

Additional information on this method is available in Reference 4.

5.4.6 Cognitive Reliability and Error Analysis Method (CREAM)

CREAM is a bidirectional analysis method that can be used for performance prediction and accident analysis. Its emphasis is on the analysis of human actions' causes (that is, human cognitive-related activities) [11]. The basis for the CREAM is the categorization schemes of error modes as well as of various components of the organization, technology, and human triad, which incorporate human-related factors, technology-related factors, and organization-related factors.

The objective of utilizing this methodology is to assist all involved analysts in the following areas [28]:

- **Area A:** Highlight all the surrounding environments in which the cognition of these conditions may be reduced and thus determine what type of actions may lead to a probable risk.

- **Area B:** Highlight tasks, work, or activities within the framework of the system boundaries which necessitate or essentially depend on a range of human thinking and which are therefore very much vulnerable to variations in their reliability level.

- **Area C:** Make essential recommendations/suggestions as to how highlighted error-generating conditions may be improved to a certain extent and how the system reliability can be improved while also decreasing risk.

- **Area D:** Compile an evaluation from the assessment of the various types of human performance–associated outcomes as well as their direct or indirect effect on system safety. In turn, this can be utilized as part of the probability risk assessment.

Four generic cognitive function failure types utilized in CREAM are shown in Figure 5.2 [12].

"Observation errors" include items such as incorrect object observation (i.e., a response is given to the incorrect event or stimulus) and observation overlooked (i.e., overlooking a measurement or a signal). "Planning errors" include items such as priority error (i.e., choosing the incorrect goal) and improper plan formulated (i.e., plan is either directly incorrect or incomplete). "Execution errors" include items such as action carried out at wrong time (i.e., either too early or too late), action overlooked, action carried out of sequence, and action of wrong type carried out with regard to speed, force, distance, or direction.

Finally, "Interpretation errors" include items such as delayed interpretation (i.e., not made in time), faulty diagnosis (i.e., either an incomplete diagnosis or a wrong diagnosis), and decision error (i.e., either not making a decision at all or making an incorrect or incomplete decision).

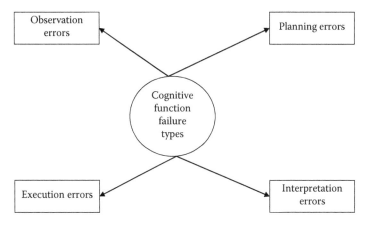

FIGURE 5.2
Generic cognitive function failure types utilized in CREAM.

Some of the benefits and drawbacks of this method (i.e., CREAM) are as follows [29]:

Benefits

- It allows for the direct human error probability quantification.
- It makes use of the same principles for predictive and retrospective analyses.
- It is quite well structured, concise, and follows a quite well laid out system of procedure.

Drawbacks

- It fails to put forth effective potential means by which the high-lighted errors can be reduced.
- It needs a rather high level of resource use, including quite lengthy time periods for completion.

Additional information on this method is available in Reference 4.

5.4.7 Standardized Plant Analysis Risk–Human Reliability Analysis (SPAR-HRA)

This method was developed at the Idaho National Laboratory for estimating the probability that an operator will fail when tasked with conducting a basis event [5,15,30]. As per References 4 and 30, the method does the following:

- Decomposes probability into contributions from diagnosis-associated failures and action-associated failures.
- Uses a beta distribution for uncertainty analysis.
- Accounts for the context associated with human failure-related events by employing performance-shaping factors and dependency assignment for adjusting a base-case human error probability.
- Makes use of designated worksheets to ensure consistency of analyst.
- Makes use of predefined and base-case human error probabilities as well as performance-shaping factors along with guidance on how to assign the right value of the performance-shaping factor.

In addition, SPAR-HRA assigns human activity to one of two general task classifications: diagnosis or action. Diagnosis tasks consist of reliance on knowledge and experience for comprehending existing situations, determining correct courses of action, and planning and prioritizing activities. Some examples of action tasks are starting machines/pumps, conducting calibration or testing, and operating equipment.

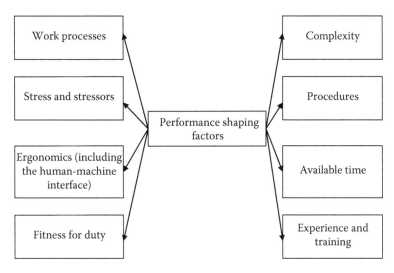

FIGURE 5.3
Performance shaping factors considered in the quantification process of SPAR-HRA.

The eight performance-shaping factors that are accounted for in the quantification process of SPAR-HRA are shown in Figure 5.3 [4,7,9,30].

The worksheet is the main element of SPAR-HRA as it significantly simplifies the estimation procedure. However, the worksheet using process varies slightly depending on whether the analyst is utilizing the method (i.e., SPAR-HRA) to carry out a more detailed HRA, conduct event analysis, or build SPAR models.

Some of the benefits and drawbacks of SPAR-HRA are as follows [4,7,31]:

Benefits
- A simple underlying model makes it relatively straightforward to use, and its end results are traceable.
- The THERP-like dependence model can be utilized for addressing both subtask as well as event sequence dependence.
- The eight performance-shaping factors included in it cover many situations in which more detailed analysis is not necessary.

Drawbacks
- For detailed analysis, the degree of resolution of the performance-shaping factors may be insufficient.
- It may not be appropriate in situations in which more detailed and realistic analysis of diagnosis-related errors is required.

Additional information on this method is available in Reference 4.

5.5 Problems

1. Write an essay on human reliability methods for nuclear power stations.
2. Discuss HRA process steps and their end results with respect to nuclear power plants.
3. What are the issues concerning the incorporation of HRA integrally into a PRA? Discuss each of these issues in detail.
4. What are the seven methods considered quite useful for performing HRA in nuclear power stations?
5. What are the information requirements for the following three HRA methods?:
   a. ASEP
   b. SLIM-MAUD
   c. THERP
6. What are the main benefits and drawbacks of the THERP method?
7. Describe the SLIM-MAUD method.
8. What were the main reasons for developing ATHEANA?
9. What are the benefits and drawbacks of the following two methods?
   a. HEART
   b. ATHEANA
10. What is CREAM? Describe it in detail.

References

1. Facts and Figures, *Nuclear Industry Association (NIA)*, Carlton House, London, 2013.
2. Ryan, T.G., A task analysis-linked approach for integrating the human factor in reliability assessments of nuclear power plants, *Reliability Engineering and System Safety*, Vol. 22, 1988, pp. 219–234.
3. Trager, T.A., *Case Study Report on Loss of Safety Function Events*, Report No. AEOD/C504, U.S. Nuclear Regulatory Commission, Washington, DC, 1985.
4. Bell, J., Holroyd, J., *Review of Human Reliability Assessment Methods*, Research Report No. RR679, Health and Safety Executive, Buxton, U.K., 2009.
5. Boring, R.L., Gertman, D.I., Joe, J.C., Marble, J.L., Human reliability analysis in the U.S. nuclear power industry: A comparison of atomistic and holistic methods, *Proceedings of the Human Factors and Ergonomics Society 49th Annual Meeting*, 2005, pp. 1815–1819.

6. Sundaramurthi, R., Smidts, C., Human reliability modeling for the next genera-tion system code, *Annals of Nuclear Energy*, Vol. 52, 2013, pp. 137–156.
7. Dhillon, B.S., *Human Reliability, Error, and Human Factors in Power Generation*, Springer, London, 2014.
8. *IEEE Guide for Incorporating Human Action Reliability Analysis for Nuclear Power Generating Stations*, IEEE Std. 1082-1997, Institute of Electrical and Electronics Engineers (IEEE), New York, 1997.
9. Spurgin, A.J., Lydell, B., Critique of current human reliability analysis methods, *Proceedings of the IEEE Seventh Conference on Human Factors and Power Plants*, 2002, pp. 3.12–3.18.
10. Dhillon, B.S., *Human Reliability: with Human Factors*, Pergamon Press, New York, 1986.
11. Strater, O., Bubb, H., Assessment of human reliability based on evaluation of plant experience: Requirements and implementation, *Reliability Engineering and Systems Safety*, Vol. 63, 1999, pp. 199–219.
12. Kim, I.S., Human reliability analysis in the man-machine interface design review, *Annals of Nuclear Energy*, Vol. 28, 2001, pp. 1069–1081.
13. Laux, L., Plott, C., Using operator workload data to inform human reliability analyses, *Proceedings of the Joint 8th IEEE HFPP/13th HPRCT Conference*, 2007, pp. 309–312.
14. Ryan, T.G., A task analysis-linked approach for integrating the human factor in reliability assessments of nuclear power plants, *Reliability Engineering and System Safety*, Vol. 22, 1988, pp. 219–234.
15. Dhillon, B.S., *Safety and Human Error in Engineering Systems*, CRC Press, Boca Raton, Florida, 2013.
16. *Human Error Classification and Data Collection*, Report No. IAEA.TECDOC 538, International Atomic Energy Agency (IAEA), Vienna, Austria, 1990.
17. Embrey, D.E. et al., *SLIM-MAUD: An Approach to Assessing Human Error Probabilities Using Unstructured Expert Judgement, Volume I: Overview of SLIM-MAUD*, Report No. NUREG/CR-3518, U.S. Nuclear Regulatory Commission, Washington, DC, 1987.
18. Humphreys, P., ed., *Human Reliability Assessor's Guide, REA Technology*, London, UK, 1995.
19. Swain, A.D., Guttman, H.E., *Handbook of Human Reliability Analysis with Emphasis on Nuclear Power Plant Applications*, Report No. NUREG/CR-1278, U.S. Nuclear Regulatory Commission, Washington, DC, 1983.
20. Kirwin, B., *A Guide to Practical Human Reliability Assessment*, CRC Press, Boca Raton, Florida, 1994.
21. Kirwan, B. et al., The validation of three human reliability quantification tech-niques, THERP, HEART, and JHEDI: Part II—Results of validation exercise, *Applied Ergonomics*, Vol. 28, No. 1, 1997, pp. 17–25.
22. Swain, A.D., *Accident Sequence Evaluation Program Human Reliability Analysis Procedure*, Report No. NUREG/CR-4772, U.S. Nuclear Regulatory Commission, Washington, DC, 1987.
23. Cooper, S.E. et al., *A Technique for Human Event Analysis (ATHEANA)-Technical Basis and Methodological Description*, Report No. NUREG/CR-6350, U.S. Nuclear Regulatory Commission, Brookhaven National Laboratory, Upton, New York, April 1996.

24. *Technical Basis and Implementation Guidelines for a Technique for Human Event Analysis (ATHEANA)*, Report No. NUREG-1624, U.S. Nuclear Regulatory Commission, Washington, DC, May 2000.
25. Forester, J. et al., Expert elicitation approach for performing ATHEANA quantification, *Reliability Engineering and System Safety*, Vol. 83, 2004, pp. 207–220.
26. Kirwan, B. et al., The validation of three human reliability quantification techniques, THERP, HEART, and JHEDI: Part III-practical aspects of the usage of the techniques, *Applied Ergonomics*, Vol. 28, No. 1, 1997, pp. 27–39.
27. Kirwin, B., The validation of three human reliability quantification techniques, THERP, HEART, and JHEDI: Part I-techniques descriptions and validation issues, *Applied Ergonomics*, Vol. 27, No. 6, 1996, pp. 359–373.
28. Hollnagel, E., *Cognitive Reliability and Error Analysis Method-CREAM*, Elsevier Science, Oxford, UK, 1998.
29. Stanton, N.A. et al., *Human Factors Methods: A Practical Guide for Engineering and Design*, Ashgate Publishing, Aldershot, UK, 2005.
30. Gertman, D. et al., *The SPAR-H Human Reliability Method*, Report No. NUREG/CR-6883, U.S. Nuclear Regulatory Commission, Washington, DC, 2004.
31. Forester, J. et al., *Evaluation of Analysis Methods Against Good Practices*, Report No. NUREG-1842, U.S. Nuclear Regulatory Commission, Washington, DC, 2006.

6

Safety in Nuclear Power Plants

6.1 Introduction

The International Atomic Energy Agency (IAEA) defines nuclear safety as "The achievement of proper operating conditions, prevention of accidents or mitigation of accident consequences, resulting in protection of workers, the public and the environment from undue radiation hazards" [1,2]. This covers nuclear power plants and all other nuclear-related facilities and areas.

Over the years, due to the occurrence of various types of accidents in nuclear power plants, safety in nuclear power plants has become a very important issue, and it requires a continuing quest for excellence. All involved organizations and individuals need constantly to be alert to opportunities for lowering risks to the lowest practical level as possible. The nuclear power industrial sector has considerably improved the safety of reactors and other related systems and has proposed safer reactor and related systems designs. However, it is not possible to guarantee perfect safety because of potential sources of problems such as human errors and external events that can have a much greater impact than anticipated.

This chapter presents various important aspects of safety in nuclear power plants.

6.2 Nuclear Power Plant Safety Objectives

As per IAEA, three safety objectives for nuclear power plants are as follows [3]:

- **General nuclear safety objective.** This objective is concerned with protecting society, individuals, and the environment by establishing and maintaining in all nuclear power plants a highly effective defense against radiological-related hazards.
- **Radiation protection objective.** This objective is concerned with ensuring in normal operation that radiation exposure within the

plant boundary and due to any release of radioactive material from the plant is as low as possible, economic and social-related factors being taken into consideration properly, and ensuring proper mitigation of the extent of radiation-related exposure due to accidents.

- **Technical safety objective.** This objective is concerned with preventing with high confidence the occurrence of accidents in nuclear plants; ensuring that the occurrence of all accidents is properly taken into account during the design process of the plant, even those with very low occurrence probability and whose radiological consequences, if any, would be minor; and ensuring that the probability of occurrence of severe accidents with serious radiological consequences is extremely low.

It should be noted that the first objective is quite general in nature, and the other two are complimentary objectives that interpret the general objective, dealing with radiation-related protection and technical aspects of safety, respectively. All in all, the safety objectives are not totally independent; their overlap adds emphasis as well as ensures completeness.

6.3 Nuclear Power Plant Fundamental Safety Principles

There are three types of nuclear power plant fundamental safety principles relating to management, defense in depth, and technical issues. Each of these types is presented below, separately [3].

6.3.1 Management Responsibilities

There are three fundamental management principles that are concerned with the establishment of a safety culture, the operating organization's responsibilities, and the provision of regulatory control and verification of safety-associated activities. Each of these items is described below [3].

- **Safety culture. In this case principle:** An established safety culture governs the interactions of all individuals and organizations involved in activities concerning nuclear power.
- **Responsibility of the operating organization. In this case principle:** The ultimate responsibility concerning the safety of a nuclear power plant totally rests with the operating organization, and this is in no way diluted by the designers', contractors', suppliers', constructors', and regulators' separate responsibilities and activities.
- **Regulatory control and independent verification. In this case principle:** The government establishes an appropriate legal framework

for the nuclear industrial sector and an impendent regulatory body for licensing and regulatory control of nuclear power plants and for enforcing the appropriate regulations. Furthermore, it is essential that the separation between the responsibilities of the regulatory body and those of other involved parties is completely clear, so that the involved regulators retain their total independence as a safety authority and are absolutely protected from undue pressure.

6.3.2 Strategy of Defense in Depth

"Defense in depth" is singled out among the basic/fundamental principles since it underlies the nuclear power plant's safety technology. Two corollary principles of "defense in depth" are accident prevention and accident mitigation. The general statement of "defense in depth," accident prevention, and accident mitigation are described below [3].

- **Defense in depth. In this case principle:** To compensate for potential human- and mechanical-related failures, an appropriate defense in depth concept is implemented, centered on various levels of protection including successive barriers to prevent the release of radioactive material to the surrounding environment.

    The concept incorporates protection of barriers by averting damage to the plant as well as to the barriers themselves. Furthermore, it also incorporates further measures for protecting the public as well as the environment from harm in case the barriers in question are not totally effective.

- **Accident prevention. In this case principle:** Principal emphasis is placed on the primary means of achieving safety effectively, which is the prevention of the occurrence of accidents, particularly those which could lead to severe core damage.

- **Accident mitigation. In this case principle:** Off-site and in-plant appropriate measures are available and are prepared, for that would considerably decrease the effects of radioactive material's accidental release.

6.3.3 General Technical Principles

There are many underlying technical principles concerning technical issues that are important to the successful application of safety technology in nuclear power plants. Seven of these principles are presented below [3].

- **Proven engineering practices. In this case principle:** Nuclear power technology is based on engineering-related practices that are clearly proven by testing and experience, and which are clearly reflected in approved standards and codes as well as in other documented statements.

- **Quality assurance. In this case principle:** Quality assurance is appropriately applied throughout activities at a nuclear power plant as an element of a comprehensive system to ensure with high confidence that all items delivered, services, and tasks carried out satisfy stated requirements.

- **Self-assessment. In this case principle:** Self-assessment for all important activities at a nuclear power plant ensures the involvement of individuals conducting line functions in detecting problems that concern safety and performance and overcoming them.

- **Peer reviews. In this case principle:** Independent peer reviews provide access to practices and programs used at power plants performing well and allow their usage at other nuclear power plants.

- **Human factors. In this case principle:** Individuals engaged in activities bearing on nuclear power plant safety are trained and qualified for carrying out their duties. The probability/possibility of the occurrence of human error in nuclear power plant operation is taken into consideration by facilitating right decisions by involved operators and inhibiting incorrect decisions and by providing appropriate mechanisms to detect and correct or compensate for error.

- **Safety assessment and verification. In this case principle:** Safety assessment is carried out prior to the start of a nuclear plant's construction and operation. The assessment is well documented and reviewed independently. This assessment is subsequently updated in the light of a significant amount of new safety-related information.

- **Radiation protection. In this case principle:** A system of radiation protection-related practices, clearly consistent with recommendations of the IAEA and the International Commission on Radiological Protection (ICRP), is followed during the design, commissioning, operational, and decommissioning phases of nuclear power plants.

6.4 Nuclear Power Plant Specific Safety Principles

Nuclear power plant safety objectives and fundamental safety principles provide a good conceptual framework for the specific safety principles. The specific safety principles are concerned with the nuclear power plant siting, design, manufacturing and construction, commissioning, accident management, decommissioning, and emergency preparedness. The safety principles concerned with each of these areas are presented below, separately [3].

6.4.1 Siting

The site is the area within which a nuclear power plant is situated and is under the full control of the operating company/organization. The selection of an appropriate site for the nuclear power plant is a very important process as local circumstances can directly or indirectly affect safety considerably.

The safety principles that are concerned with the external factors affecting the plant, radiological impact on the public and the local environment, feasibility of emergency plans, and ultimate heat sink provisions are as follows, respectively [3]:

- **External factors affecting the plant. In this case principle:** The selection of the site clearly takes into consideration the findings of investigations of local factors that can adversely affect the power plant safety.
- **Radiological impact on the public and the local environment. In this case principle:** The sites under consideration are investigated from the standpoint of the radiological impact of the power plant during normal operation and in accident conditions.
- **Feasibility of emergency plans. In this case principle:** The site chosen for a nuclear power plant is quite compatible with the off-site countermeasures that may be necessary for limiting the effects of the radioactive substances' accidental releases and is expected to remain quite compatible with such measures.
- **Ultimate heat sink provisions. In this case principle:** The site chosen for a nuclear power plant contains a highly reliable long-term heat sink that can totally remove energy produced in the power plant after shutdown, both right after shutdown as well over the longer period.

6.4.2 Design

The primary objective of designers of a nuclear power plant is providing a good design. Those individuals ensure that the power plant's structures, systems, and components have the proper characteristics, material composition and specifications, and are integrated and laid out in such a way as to satisfy the general plant performance-related specifications effectively.

The design-related safety principles can be grouped under three areas: design process, general features, and specific features. The design process-related safety principles are concerned with design management, proven technology, and general basis for design. Similarly, the general features-related safety principles are concerned with plant process control systems, automatic safety systems, reliability targets, dependent failures, equipment qualification, inspectability of safety equipment, and radiation protection in design.

Finally, the specification features-related safety principles are concerned with protection against power-transient accidents; reactor core integrity; automatic shutdown systems; normal heat removal; startup, shutdown, and low power operation; emergency heat removal; reactor coolant system integrity; confinement of radioactive material; protection of confinement structure; monitoring of plant safety status; preservation of control capability; station blackout; control of accidents within the design basis; new and spent fuel storage; and plant physical protection.

Additional information on design-related safety principles is available in Reference 3.

6.4.3 Manufacturing and Construction

In this case, a primary safety requirement is that a nuclear power station be manufactured and constructed as per the design intent. This is achieved by maintaining appropriate attention to a variety of issues, from the broad aspect of accountability of all involved organizations to the competence, diligence, and proper care of the individual personnel/workers.

The safety principles that are concerned with the safety evaluation of design and the achievement of quality are as follows [3]:

- **Safety evaluation of design. In this case principle:** Construction of a nuclear power station is started only after the regulatory body and the operating company/organization have fully satisfied themselves through appropriate assessments that the major safety-related issues have been resolved and that the remainder are amendable to solution prior to when operations are scheduled to start.

- **Achievement of quality. In this case principle:** The power plant constructors and manufacturers discharge their responsibilities for the provision of equipment and construction of good quality by applying well-proven and developed methods and procedures clearly supported by quality assurance-related practices.

6.4.4 Commissioning

Commissioning is necessary for demonstrating that the completed power plant is satisfactory for service prior to it being made operational. For this very purpose, a well-planned and well-documented commissioning program is developed and executed. The operating company/organization, including its potential operating staff personnel, take part in this phase. Power plant systems are progressively passed over to the operating staff personnel as the installation and testing of each and every item are accomplished.

The safety principles that are concerned with verification of design and construction, validation of operating and functional test procedures, collecting baseline data, and preoperational plant adjustments are as follows [3]:

- **Verification of design and construction. In this case principle:** The commissioning program is developed and followed for demonstrating that the whole power plant, particularly items critical to safety and radiation protection, has been constructed and operates as per the design intent, and for ensuring that weaknesses are detected and rectified.

- **Validation of operating and functional test procedures. In this case principle:** All procedures involved for normal plant and systems operation and for functional tests to be carried out during the operational phase are validated as part of the commissioning program.

- **Collecting baseline data. In this case principle:** During commissioning-related tests, detailed diagnostic-related data are collected on parts that have special safety-related significance, and the system's initial operating parameters are documented.

- **Preoperational plant adjustments. In this case principle:** During the commissioning process, the as-built safety and process systems operating characteristics are determined and recorded. All operating points are adjusted for conforming to design-related values as well as to safety analyses. All training-related procedures and limiting conditions for operation are changed for reflecting accurately the systems as-built operating characteristics.

6.4.5 Operation

The operating company/organization is fully responsible to provide all equipment, staff, procedures, and management practices appropriate for safe operation, including the fostering of an effective environment in which safety is clearly seen as an important factor and a matter of personal accountability for all involved staff members. It may seem on certain occasion that emphasis on safety might be in conflict with the requirement for accomplishing a high-capacity factor and for satisfying all demands of electricity generation. This very conflict is rather more apparent than real as well as it can at most be transitory.

The operation-related safety principles are concerned with organization, responsibility, and staffing; safety review procedures; conduct of operations; training; operational limits and conditions; normal operating procedures; emergency operating procedures; radiation protection procedures; engineering and technical support of operations; feedback of operating experience; maintenance, testing, and inspection; and quality assurance in operation.

Additional information on operation-related safety principles is available in Reference 3.

6.4.6 Accident Management

Accident management as an element of accident prevention includes the measures to be carried out by operators during an accident sequence's evolution after conditions have come to exceed the plant's design but prior to the development of a severe accident. Similarly, accident management as an element of accident mitigation includes constructive measure by the operating staff members in the event of the occurrence of a severe accident, clearly directed to preventing the further progress of such an accident as well as alleviating its affects.

The safety principles that are concerned with strategy for accident management, training and procedures for accident management, and engineered features for accident management are as follows [3]:

- **Strategy for accident management. In this case principle:** The findings of an analysis of the response of the power plant to future accidents beyond the design basis are used appropriately in developing guidance on an accident management-related strategy.
- **Training and procedures for accident management. In this case principle:** All nuclear power plant staff members are appropriately trained and retrained in the procedures to be followed if an accident occurs that exceeds the plant's design basis.
- **Engineered features for accident management. In this case principle:** All equipment, instrumentation, and diagnostic aids are available to all involved operators, who may at some time be faced with the need for controlling the consequences and course of an accident that are beyond the design basis.

6.4.7 Decommissioning

A nuclear power plant that is shut down remains an operating plant until its decommissioning takes place and is subject to normal control procedures and processes for ensuring safety. After the termination of operations and the removal of spent fuel from the power plant, a significant radiation hazard still remains, which must be managed properly for protecting the workers' health and the public.

The removal of power plant equipment and its decontamination can only be facilitated effectively if proper consideration is given during the design phase to decommissioning and disposal of the wastes arising from the decommissioning process. The safety principle concerned with the decommissioning is as follows [3]:

- **Principle:** Proper consideration is given during design and plant operations for facilitating eventual decommissioning and waste management. After the termination of operations and the spent fuel's removal from the plant, radiation-related hazards are managed so as to protect all involved workers' health and the public during the plant decommissioning process.

6.4.8 Emergency Preparedness

Emergency planning and preparedness is comprised of actions required for ensuring that, in the event of the occurrence of an accident, all measures required for protecting the public and the staff members could be performed and that the use of such services would be disciplined properly.

The safety principles that are concerned with emergency plans, emergency response facilities, and assessment of accident consequences and radiological monitoring are as follows [3]:

- **Emergency plans. In this case principle:** Appropriate emergency plans are developed prior to the start-up of the power plant and are exercised periodically for ensuring that all protection-related actions can be implemented in the event of an accident occurrence which results in (or has the potential for) quite significant releases of radioactive-related material within as well as beyond the site boundary. All emergency planning zones defined around the power plant clearly allow for the application of a graded response.

- **Emergency response facilities. In this case principle:** For emergency response, a permanently equipped emergency center off the site is available. Also, a similar center on the site is provided to direct emergency-related actions within the power plant and to communicate with the off-site emergency organization/agency.

- **Assessment of accident consequences and radiological monitoring. In this case principle:** All means are clearly available to the responsible site staff members to be employed in early prediction of the extent as well as significance of any release of radioactive-related materials if an accident were to take place, for quick and continuous assessment of the radiological condition, and for determining the necessity for protective actions.

6.5 Management of Safety in Nuclear Power Plant Design

The management of safety in nuclear power plant design is very important. In this regard, the requirements are as follows [4–6]:

- **Requirement A: Responsibilities in the management of safety in power plant design.** In this case, an applicant for a licence for constructing and/or operating a nuclear power plant shall be fully responsible to ensure that the design submitted to the regulatory organization/body clearly satisfies each and every applicable safety-related requirement.

- **Requirement B: Management system for power plant design.** In this case, the design organization shall develop and implement a management system to ensure that each and every safety-related requirement developed for the design of the power plant is considered and implemented properly in all phases of the design process, and all these requirements are met effectively in the final design.

- **Requirement C: Safety of the power plant design throughout the lifetime of the power plant.** In this case, the power plant's operating organization shall develop a formal system to ensure the continuing safety of the power plant design throughout the nuclear power plant's lifetime.

Additional information on the above three requirements is available in Reference 4.

6.6 Safety-Related Requirements in the Design of Specific Nuclear Plant Systems

These requirements may be grouped under the following 10 areas:

- **Area 1: Reactor core and associated features.** In this area, the requirements are concerned with performance of fuel elements and assemblies, structural capability of the reactor core, control of the reactor core, and reactor shutdown.

- **Area 2: Reactor coolant systems.** In this area, the requirements are concerned with the design of reactor coolant systems, overpressure protection of the reactor coolant pressure boundary, inventory of reactor coolant, cleanup of reactor coolant, removal of residual heat from the reactor core, emergency cooling of the reactor core, and heat transfer to an ultimate heat sink.

- **Area 3: Containment structure and containment system.** In this area, the requirements are concerned with the containment system for the reactor, control of radioactive releases from the containment, isolation of the containment, access to the containment, and control of containment conditions.

- **Area 4: Instrumentation and control systems.** In this area, the requirements are concerned with provision of instrumentation, control systems, protection systems, reliability and testability of instrumentation and control systems, use of computer-based equipment in systems important to safety, separation of protection systems and control systems, control room, supplementary control room, and emergency control center.

- **Area 5: Emergency power supply.** In this area, there is only one requirement, and it is totally concerned with emergency power supply.
- **Area 6: Supporting systems and auxiliary systems.** In this area, the requirements are concerned with performance of supporting systems and auxiliary systems, heat transport systems, process sampling systems and postaccident sampling systems, compressed air systems, air conditioning systems and ventilation systems, fire protection systems, lighting systems, and overhead lifting equipment.
- **Area 7: Other power conversion systems.** In this area, there is only one requirement, and it is concerned with steam supply system, feedwater system, and turbine generators.
- **Area 8: Treatment of radioactive effluents and radioactive waste.** In this area, the requirements are concerned with systems for treatment and control of waste and systems for treatment and control of effluents.
- **Area 9: Fuel handling and storage systems.** In this area, there is only one requirement, and it is concerned totally with fuel handling and storage systems.
- **Area 10: Radiation protection.** In this area, the requirements are concerned with design for radiation protection and means of radiation monitoring.

Ten of the above 10 areas' requirements are presented below [4]:

- **Requirement 1:** This requirement is concerned with reactor shutdown, and it is that the appropriate means shall be provided for ensuring that there is an effective capability for shutting down the nuclear plant's reactor in operational states and in accident situations and that the shutdown condition can be kept even for the reactor core's most reactive conditions.
- **Requirement 2:** This requirement is concerned with the cleanup of reactor coolant, and it is that the appropriate facilities shall be allocated at the nuclear power plant to remove radioactive substances from the reactor coolant, including activated corrosion-related products and fission-related products derived from the fuel and all nonradioactive substances.

- **Requirement 3:** This requirement is concerned with access to the containment and it is that access by operating staff members to the containment at a nuclear power facility shall be through airlocks appropriately equipped with doors that are interlocked to ensure that at least one of the doors is closed during reactor power operation phase as well as in accident situations.

- **Requirement 4:** This requirement is concerned with reliability and testability of instrumentation and control systems, and it is that at the nuclear power facility/plant, instrumentation and control systems for equipment/items important to safety shall be designed for very high reliability and periodic testability commensurate with the safety-related function(s) to be conducted.

- **Requirement 5:** This requirement is concerned with emergency power supply, and it is that at the nuclear power facility/plant, the emergency power supply shall be capable of supplying the required power in expected operational occurrences and accident situations in the event of the off-site power loss.

- **Requirement 6:** This requirement is concerned with heat transport systems, and it is that auxiliary systems shall be provided as required for removing heat from systems and parts at the nuclear power facility/plant that are essential to function in operational states as well as in accident situations.

- **Requirement 7:** This requirement is concerned with lighting systems, and it is that adequate lighting shall be provided in all of a nuclear power facility's/plant's operational areas in operational states as well in accident situations.

- **Requirement 8:** This requirement is concerned with systems for treatment and control of effluents, and it is that appropriate systems shall be provided at the nuclear power facility/plant to treat liquid and gaseous radioactive effluents for keeping their amounts well below the authorized limits on discharges and as low as reasonably possible to achieve.

- **Requirement 9:** This requirement is concerned with fuel handling and storage systems, and it is that appropriate fuel handling and storage systems shall be provided at the nuclear power facility/plant to ensure that the fuel's integrity and properties are effectively maintained at all times during the fuel handling and storage process.

- **Requirement 10:** This requirement is concerned with the means of radiation monitoring, and it is that the appropriate equipment shall be provided at the nuclear power facility/plant to ensure that there is appropriate radiation monitoring in operational states and design basis accident situations and, as far as is possible, in design extension situations.

6.7 Deterministic Safety Analysis for Nuclear Power Plants

Throughout the lifetime of a nuclear power plant, safety analyses play an instrumental role. The stages of and occasions in the lifetime of a nuclear power plant in which the application of safety analyses is relevant include the following [7]:

- Design
- Commissioning
- Operation and shutdown
- Modification of operation or design
- Periodic safety-related review
- Life extension in states/provinces where licences are issued for a limited period

Two basic types of safety analysis are deterministic safety analysis and probabilistic safety analysis. Deterministic safety analyses for a nuclear power plant/facility predict the response to postulated initiating events, and a specific set of rules and acceptance criteria is applied. Usually, they focus on radiological, neutronic, thermohydraulic, thermomechanical, and structural aspects, which are frequently analyzed with different computational methods. The computations are normally performed for predetermined operational states and operating modes, and the events incorporate expected postulated accidents, selected beyond design basis accidents, severe accidents with core degradation, and transients.

The options for performing deterministic safety analyses are shown in Figure 6.1 [7].

Additional information on Figure 6.1 options is available in Reference 7.

6.7.1 Deterministic Safety Analysis Application Areas

There are six areas in which deterministic safety analysis can be applied in nuclear power plants. These areas are as follows [7]:

- **Area 1: Nuclear power plant's design.** In this case, deterministic safety analyses need either a conservative approach or a clearly best estimate analysis together with an evaluation of all uncertainties.

- **Area 2: Analysis of incidents that have taken place or of combinations of such incidents with all other hypothetical faults.** In this case, analyses would generally require best estimate approaches, in particular for sophisticated occurrences that need a realistic simulation.

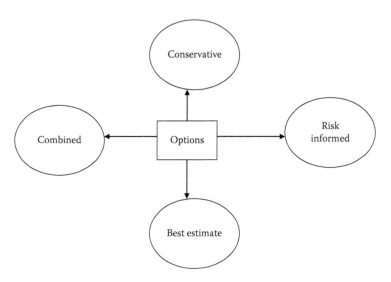

FIGURE 6.1
Options for performing deterministic safety analyses in nuclear power plants.

- **Area 3: Production of new or revised safety analysis documents for licensing, including obtaining the approval of the regulatory organization/body for changes to a power plant and to power plant operation.** In this case, both conservative methods and best estimate plus uncertainty approaches may be employed.

- **Area 4: Development and maintenance of emergency operating-related procedures and accident management-related procedures.** In this case, best estimate codes along with realistic assumptions should be used.

- **Area 5: Assessment by the regulatory body/organization of safety analysis-related reports.** In this case, both conservative methods and best estimate plus uncertainty approaches may be employed.

- **Area 6: Refinement of earlier safety-related analyses in the context of a periodic safety review for providing assurance that the earlier conclusions and assessments are still valid.**

6.8 Nuclear Power Plant Safety-Related Documents and Standards

Over the years, the IAEA has produced many documents/standards directly or indirectly concerned with nuclear power plant safety. Some of these documents/standards are as follows [8]:

- Fundamental Safety Principles, Series No. SF-1, International Atomic Energy Agency, Vienna, Austria, 2006
- Fire Safety in the Operation of Nuclear Power Plants, Series No. NS-G-2.1, International Atomic Energy Agency, Vienna Austria, 2000.
- Severe Accident Management Programmes for Nuclear Power Plants, Series No. NS-G-2.15, International Atomic Energy Agency, Vienna, Austria, 2009.
- Evaluation of Seismic Safety for Existing Nuclear Installations, Series No. NS-G-2.13, International Atomic Energy Agency, Vienna, Austria, 2009.
- Radiation Protection Aspects of Design for Nuclear Power Plants, Series No. NS-G-1.13, International Atomic Energy Agency, Vienna, Austria, 2005.
- Safety of Nuclear Fuel Cycle Facilities, Series No. NS-R-5 (Rev. 1), International Atomic Energy Agency, Vienna, Austria, 2014.
- Establishing the Safety Infrastructure for a Nuclear Power Programme, Series No. SSG-16, International Atomic Energy Agency, Vienna, Austria, 2012.
- Periodic Safety Review for Nuclear Power Plants, Series No. SSG-25, International Atomic Energy Agency, Vienna, Austria, 2013.
- Disposal of Radioactive Waste, Series No. SSR-5, International Atomic Energy Agency, Vienna, Austria, 2011.
- Regulatory Control of Radioactive Discharges to the Environment, Series No. WS-G-2.3, International Atomic Energy Agency, Vienna, Austria, 2000.
- Decommissioning of Nuclear Power Plants and Research Reactors, Series No. WS-G-2.1, International Atomic Energy Agency, Vienna, Austria, 1999.
- Organization and staffing of the Regulatory Body for Nuclear Facilities, Series No. GS-G-1.1, International Atomic Energy Agency, Vienna, Austria, 2002.
- The Management System for Nuclear Installations, Series No. GS-G-3.5, International Atomic Energy Agency, Vienna, Austria, 2009.
- Design of Fuel Handling and Storage Systems in Nuclear Power Plants, Series No. NS-G-1.4, International Atomic Energy Agency, Vienna, Austria, 2003.
- Design of Reactor Containment Systems for Nuclear Power Plants, Series No. NS-G-1.10, International Atomic Energy Agency, Vienna, Austria, 2004.
- Seismic Design and Qualification for Nuclear Power Plants, Series No. NS-G-1.6, International Atomic Energy Agency, Vienna, Austria, 2003.

- Maintenance, Surveillance and In-Service Inspection in Nuclear Power Plants, Series No. NS-G-2.6, International Atomic Energy Agency, Vienna, Austria, 2002.
- Modifications to Nuclear Power Plants, Series No. NS-G-2.3, International Atomic Energy Agency, Vienna, Austria, 2001.
- Maintenance, Periodic Testing and Inspection of Research Reactors, Series No. NS-G-4.2, International Atomic Energy Agency, Vienna, Austria, 2006.
- Radiation Protection and Radioactive Waste Management in the Design and Operation of Research Reactors, Series No. NS-G-4.6, International Atomic Energy Agency, Vienna, Austria, 2009.
- Design of Electrical Power Systems for Nuclear Power Plants, Series No. SSG-34, International Atomic Energy Agency, Vienna, Austria, 2016.
- Commissioning for Nuclear Power Plants, Series No. SSG-28, International Atomic Energy Agency, Vienna, Austria, 2014.
- Safety Classification of Structures, Systems and Components in Nuclear Power Plants, Series No. SSG-30, International Atomic Energy Agency, Vienna, Austria, 2014.
- Predisposal Management of Radioactive Waste from Nuclear Fuel Cycle Facilities, Series No. SSG-41, International Atomic Energy Agency, Vienna, Austria, 2016.

Additional information on documents/standards directly or indirectly concerned with nuclear power plant safety is available in Reference 8.

6.9 Problems

1. Write an essay on safety in nuclear power plants.
2. What are the safety objectives for nuclear power plants?
3. Discuss the nuclear power plant fundamental safety principles relating to management and technical issues.
4. Discuss the nuclear power plant safety principles with regard to the following two areas:
   a. Design
   b. Commissioning
5. Compare the nuclear power plant safety principles in the area of commissioning with in the area of decommissioning.

6. What are the requirements with regard to management of safety in nuclear power plant design?

7. Discuss the safety-related requirements with the following items:

    a. Reactor shutdown

    b. Cleanup of reactor coolant

8. List at least 10 standards/documents directly or indirectly concerned with nuclear power plant safety.

9. What are the safety principles concerned with verification of design and construction and validation of operating and functional test procedures?

10. What are the nuclear power plant fundamental safety principles relating to defense in depth?

References

1. IAEA Safety Glossary: Version 2, International Atomic Energy Agency (IAEA), Vienna, Austria, 2007.
2. Nuclear Safety and Security, retrieved on September 8, 2016 from website: https://en.wikipedia.org/wiki/Nuclear-safety-and-security
3. Basic Safety Principles for Nuclear Power Plants, a report by the International Nuclear Energy Agency Advisory Group, Report No. 75-1NSAG-3 Rev.1: INSAG-12, International Atomic Energy Agency, Vienna, Austria, 1999.
4. Safety of Nuclear Power Plants: Design, Report No. SSR-2/1, International Atomic Energy Agency, Vienna, Austria, 2012.
5. The Management System for Facilities and Activities, IAEA Safety Standards Series No. GS-R-3, International Atomic Energy Agency, Vienna, Austria, 2006.
6. Maintaining the Design Integrity of Nuclear Installations Throughout Their Operating Life, Report No. INSAG-19, International Atomic Energy Agency, Vienna, Austria, 2003.
7. Deterministic Safety Analysis for Nuclear Power Plants: Specific Safety Guide, Report No. SSG-2, International Atomic Energy Agency, Vienna, Austria, 2009.
8. List of all Valid Safety Standards, retrieved on September 8, 2016 from website: http://www-ns.iaea.org/standards/documents/pubdoc-list.asp

7

Nuclear Power Plant Accidents

7.1 Introduction

Nowadays, nuclear power plant accidents have become an important issue around the globe. During the period from 1952–2009, there have been at least 99 (civilian and military) nuclear reactor-related accidents throughout the world that either resulted in loss of human life or greater than U.S. $50,000 of property damage [1,2]. These accidents involved loss of coolant, explosions, fires, and meltdowns and took place during both emergency conditions (e.g., earthquakes and droughts) and normal operation. Furthermore, the property damage costs included items such as destruction of property, court claims, environmental remediation, and emergency response [3].

In the United States, at least 56 nuclear reactor accidents have occurred, and relatively few accidents have involved fatalities [3]. The most serious of these accidents was the Three Mile Island accident that took place on March 28, 1979. Although this accident caused no fatalities, its estimated cost was $2.4 billion (in 2006 U.S. dollars) [1,2,4].

This chapter presents various important aspects of nuclear power plant accidents.

7.2 The International Nuclear Event Scale (INES)

The International Atomic Energy Agency (IAEA) introduced INES in 1990 to enable prompt communication of safety-significant information in case of nuclear accidents [5]. The INES is intended to be logarithmic, similar to the moment magnitude scale being used for describing the comparative magnitude of earthquakes. Each increasing level of the INES denotes an accident roughly 10 times more severe than the proceeding level. It should be noted that in comparison to earthquakes, for which the intensity of the event can be evaluated quantitatively, the severity level of man-made disasters such as a nuclear power plant accident is more subject to interpretation.

The INES is divided into the following levels [6]:

- **Level 7:** Major accident
- **Level 6:** Serious accident
- **Level 5:** Accident with wider consequences
- **Level 4:** Accident with local consequences
- **Level 3:** Serious incident
- **Level 2:** Incident
- **Level 1:** Anomaly
- **Level 0:** Deviation

Levels 7 and 6 consider the impact on people and the environment, and levels 5 and 4 consider the impact on people and the environment and the impact on radiological barriers and control. Similarly, levels 3 and 2 consider the impact on people and the environment, the impact on radiological barriers and control, and impact on defense-in-depth.

Finally, levels 1 and 0 consider impact on defense-in-depth and deviation (i.e., no safety significance), respectively.

Additional information on the above levels is available in Reference 6.

7.3 World Nuclear Power Plant Accidents' Fatalities, Rankings, and Costs

Over the years, many nuclear power plant accidents have occurred around the globe. The ones that caused fatalities were as follows [2]:

- **Mihama accident.** This accident occurred on August 9, 2004, in Mihama, Japan, where a steam explosion at the Mihama-3 station caused five fatalities.
- **Fukushima accident.** This accident occurred on February 22, 1993, in Fukushima, Japan, where high-pressure steam caused one fatality and two injuries.
- **Chernobyl accident.** This accident occurred on April 26, 1986, in Pripyat, Ukraine, where a steam explosion and meltdown caused around 45 fatalities directly and thousands eventually.
- **Idaho Falls accident.** This accident occurred on January 3, 1961, in Idaho Falls, Idaho, United States, where an explosion at the National Reactor Testing Station's SL-1 Stationary low-power reactor No. 1 caused three fatalities.

Over the years, many nuclear power plant accidents that have occurred around the globe have been ranked according to the INES. Some of these ranked accidents are as follows [7]:

- **Fukushima accident.** This accident occurred in 2011 in Fukushima, Japan, and as per INES, it was ranked 7. The cause of the accident was a reactor shutdown after the Sendai earthquake and tsunami and the failure of emergency cooling that resulted in an explosion.
- **Sellafield accident.** This accident occurred in 2005 in Sellafield, United Kingdom, and as per INES, it was ranked 3. The cause of the accident was the release of a large quantity of radioactive material, and it was contained within the installation.
- **Ishikawa accident.** This accident occurred in 1999 in Ishikawa, Japan, and as per INES, it was ranked 2. The cause of the accident was control rod malfunction.
- **Tomsk accident.** This accident occurred in 1993 in Tomsk, Russia, and as per INES, it was ranked 4. The cause of the accident was pressure buildup that led to an explosive mechanical failure.
- **Cadarache accident.** This accident occurred in 1993 in Cadarache, France, and as per INES, it was ranked 2. The cause of the accident was the spread of contamination to an area not expected by design.
- **Chernobyl accident.** This accident occurred in 1986 in Chernobyl, Ukraine, and as per INES, it was ranked 7. The accident had widespread health and environmental effects and caused the external release of a significant fraction of reactor core inventory.
- **Saint Laurent des Eaux accident.** This accident occurred in 1980 in Saint Laurent des Eaux, France, and as per INES, it was ranked 4. The cause of the accident was the melting of one channel of fuel in the reactor with no release outside the site.
- **Three Mile Island accident.** This accident occurred in 1979 in Three Mile Island, United States, and as per INES, it was ranked 5. The accident caused severe damage to the reactor core.
- **Kyshtym accident.** This accident occurred in 1957 in Kyshtym, Russia, and as per INES, it was ranked 6. The cause of the accident was a significant release of radioactive material into the environment from an explosion of a high-activity waste tank.
- **Windscale Pile accident.** This accident occurred in 1957 in Windscale Pile, United Kingdom, and as per INES, it was ranked 5. The cause of the accident was the release of radioactive material into the environment following a fire in a reactor core.
- **Chalk River accident.** This accident occurred in 1952 in Chalk River, Canada, and as per INES, it was ranked 5. The cause of the accident

was a reactor shutoff rod failure, combined with several operator errors, which led to a major power excursion of more than double the reactor's rated output.

The costs of nuclear power plant accidents have resulted in a vast sum of money. The estimated costs of some of these accidents that have occurred in various countries are presented below [2].

- **Saclay accident.** This accident occurred on July 25, 1979 in Saclay, France, when radioactive fluids escaped into drains designed for ordinary wastes and seeped into the local watershed at the Sanclay BL3 reactor. The estimated cost of this accident in 2006 was $5 million.
- **Blayais accident.** This accident occurred on December 27, 1999 in Blayais, France, when an unexpectedly strong storm flooded the Blayais Nuclear Power Plant and forced an emergency shutdown after injection pumps and containment safety systems failed from water damage. The estimated cost of this accident in 2006 was $55 million.
- **Greifswald accident.** This accident occurred on November 24, 1989, in Greifswald, Germany, when a near core meltdown occurred at the Greifswald Nuclear Power Plant. Three out of the plant's six cooling water pumps were turned off for a test, and the fourth pump broke down; the reactor control was lost and 10 fuel elements were damaged. The estimated cost of this accident in 2006 was $443 million.
- **Kalpakkam accident.** This accident occurred on October 22, 2002, in Kalpakkam, India, when almost 100 kilograms of radioactive sodium at a fast breeder reactor leaked into a purification cabin and ruined a number of valves and operating systems. The estimated cost of this accident in 2006 was $30 million.
- **Browns Ferry accident.** This accident occurred on March 22, 1975, in Browns Ferry, Alabama, United States, when fire burned for seven hours and damaged more than 1,600 control cables for three nuclear reactors at the power plant, disabling core cooling systems. The estimated cost of this accident in 2006 was $240 million.
- **Plymouth accident.** This accident occurred on April 11, 1986, in Plymouth, Massachusetts, United States, when recurring equipment problems forced an emergency shutdown of Boston Edison's Pilgrim Nuclear Power Plant. The estimated cost of this accident in 2006 was $1.001 billion.
- **Newport accident.** This accident occurred on December 25, 1993, in Newport, Michigan, United States. It is concerned with the shutdown

of Fermi Unit 2 after the main turbine experienced a major failure due to improper maintenance. The estimated cost of this accident in 2006 was $67 million.

- **Crystal River accident.** This accident occurred on September 2, 1996, in Crystal River, Florida, United States, when balance-of-plant equipment failure forced a shutdown and extensive repairs at Crystal River Unit 3. The estimated cost of this accident in 2006 was $384 million.

- **Braidwood accident.** This accident occurred on June 16, 2005, in Braidwood, Illinois, United States, when Exelon's Braidwood Nuclear Power Plant leaked tritium and contaminated local water supplies. The estimated cost of this accident in 2006 was $41 million.

- **Vernon accident.** This accident occurred on February 1, 2010, in Vernon, Vermont, United States, when deteriorated underground pipes from the Vermont Yankee Nuclear Power Plant leaked radioactive tritium into groundwater supplies. The estimated cost of this accident in 2006 was $700 million.

Additional information on world nuclear power plant fatalities, rankings, and costs is available in References 2 and 7.

7.4 Three Mile Island Accident

This accident occurred in the Three Mile Island Nuclear Power Plant located in Dauphin County, Pennsylvania, United States, on March 28, 1979. The accident was a partial nuclear meltdown in reactor No. 2 of the power plant, and it was the most significant accident in the history of U.S. commercial nuclear power plants [8,9]. It was rated 5 on the INES: Accident with Wider Consequences [7,8].

The accident started with failures occurring in the nonnuclear secondary system, followed by a stuck-open, pilot-operated relief valve of the primary system that allowed very large quantities of nuclear reactor coolant to escape. The mechanical failures were compounded by the power plant operators' initial failure to clearly recognize the condition as a loss-of-coolant accident due to improper training and human factors, such as human-computer interaction design oversights concerning ambiguous control room indicators in the user interface of the power plant. More clearly, a hidden indicator light led to an operator to override the reactor's automatic emergency cooling system because the operator in question wrongly believed there was too much coolant water in the reactor, which was causing the steam pressure release [8].

7.4.1 Radiological Health Effects

The accident raised concerns regarding the possibility of radiation-induced health effects, principally cancer, particularly in the area that surrounded the power plant. Due to these concerns, the Pennsylvania Department of Health maintained a registry of over 30,000 persons for 18 years who lived within a five-mile radius of Three Mile Island at the time of the occurrence of the accident [10]. In 1997, the registry was stopped, without any evidence of unusual health-related trends in the area.

All in all, over a dozen major independent health-related studies concerning the accident showed no evidence of any abnormal number of cancers around Three Mile Island, years after the occurrence of the accident.

7.4.2 The Three Mile Island Reactor No. 2 Cleanup

The cleanup of the damaged reactor system took almost 12 years, and its cost was around $973 million [9]. The cleanup was uniquely very challenging technically and radiologically, and a cleanup plan was developed and executed successfully and safely by a team of over 1000 skilled workers.

The cleanup operations generated more than 10.6 megaliters of accident-generated water, and it was processed, stored, and evaporated safely. The cleanup started in August 1979 and ended in December 1993. Additional information on the cleanup is available in Reference 9.

7.4.3 The Accident's Effect on the Nuclear Power Industry

Following this accident, each year during the period of 1980–1998, the number of nuclear reactors under construction in the United States declined. At the time of the Three Mile Island accident, 129 nuclear power plants in the United States had been approved. However, of those, only 53 (i.e., the ones which were not already operating) were completed. During the lengthy review process, the federal government requirements for correcting safety-related issues and design-related shortcomings became more stringent, costs skyrocketed, and construction times were lengthened quite significantly.

Since the year prior to the Three Mile Island accident until 2012, no U.S. nuclear power plant was authorized to start construction [8].

7.5 The Chernobyl Accident

This accident occurred in the Chernobyl Nuclear Power Plant located in the city of Pripyat, Ukraine, on April 26, 1986. On this day, the power plant's reactor No. 4 suffered a catastrophic power increase, resulting in explosions in its core. Explosions and fire released large quantities of radioactive particles into atmosphere, which spread over various parts of Europe. The accident caused between 31–45 fatalities, and around 140 persons suffered various degrees of radiation-related sickness and radiation-associated acute health impairment [11]. As per Reference 12, the long-term effects, such as cancers as a result of the accident, are still under investigation.

The cost of the Chernobyl accident was estimated to be around $6.7 billion in 2006, and it was rated 7 on the INES (i.e., the maximum classification) [2,7,12,13]. Thus, in terms of causalities and cost, the Chernobyl accident was the worst nuclear power plant accident in the world [12].

7.5.1 Dispersion and Disposition of Radionuclides

The release of radioactive materials from the Chernobyl accident into the atmosphere consisted of finely fragmented nuclear fuel particles, gases, and aerosols. This release was very high in quantity and involved a quite large fraction of the radioactive product inventory that existed in the reactor. Furthermore, the duration of the release was over a 10-day period with varying rates of the release [11].

The duration and rather high altitude (i.e., around 1 kilometer) reached by the release were basically due to the graphite fire that was quite difficult to extinguish until the 10th day. Additional information on this topic is available in Reference 11.

7.5.2 Operator Error

The operator error is considered as one of the factors in the occurrence of the accident. According to the findings of some investigations, at the time of the occurrence of the accident, the reactor was being operated with many pivotal safety systems turned off. Most notably, these systems were the local automatic control system (LAR), emergency core cooling system (ECCS), and emergency power reduction system.

As per Reference 12, the involved personnel had poor understanding of technical procedures involved with the nuclear reactor in question and ignored regulation for speeding up test completion. Additional information on this topic is available in Reference 12.

7.6 The Fukushima Accident

This accident occurred in the Fukushima Nuclear Power Plant located in Fukushima, Japan, on March 11, 2011. The accident was initiated primarily by the tsunami following the Tohoku earthquake on the same day [14,15]. Soon after the occurrence of the earthquake, the power plant's active reactors automatically turned off their sustained fission reactions. However, the tsunami completely destroyed the emergency generators cooling the reactors that in turn caused reactor No. 4 to overheat from the heat decayed by the fuel rods. Consequently, the poor cooling resulted in three nuclear meltdowns and the release of radioactive material starting on March 12, 2011. Furthermore, during the period from March 12, 2011, to March 15, 2011, several hydrogen-air chemical explosions occurred.

This accident was rated 7 on the INES (i.e., the maximum classification). It received the same ranking as the Chernobyl accident. It means that the Fukushima and Chernobyl accidents were the largest nuclear power plant disasters in the world [15].

In July 2012, the findings of the Fukushima Nuclear Accident Independent Investigation Commission (NAIIC) revealed that the causes for the occurrence of the accident had been foreseeable, and that the operator of the plant, Tokyo Electric Power Company (TEPCO), had failed to satisfy various basic safety requirements (for example, developing evacuation plans, risk assessment, and preparing for continuing collateral damage). TEPCO admitted for the first time, in October 2012, that it had failed to take appropriate actions because of fear of inviting lawsuits or protests against its nuclear power plants [15,16].

Additional information on Fukushima accident is available in References 14–17.

7.7 Lessons Learned from the Three Mile Island, Chernobyl, and Fukushima Accidents

There are many lessons that could be learned from the Three Mile Island, Chernobyl, and Fukushima accidents. For the Three Mile Island accident, the nonavailability of the emergency feed system, the erroneous signals generated by the instrumentation with respect to the pressure vessel's "open" valves, poor knowledge of the power plant's real status, and the primary pumps' shutdown resulted in a partial core meltdown [18].

Similarly, with regard to the Chernobyl accident, some of the lessons learned are as follows [19]:

- The accident has clearly demonstrated the definite need for establishing and supporting a high-level national emergency response system in case of man-made accidents.
- The accident has demonstrated that the nuclear facilities' safety ensuring cost is significantly less than that of dealing with the consequences of the accident.

- The accident has clearly demonstrated the importance of strict compliance with the fundamental and technical principles of safety for nuclear power plants.
- The scale of material losses and the Chernobyl accident's consequences mitigating financial cost provide compelling evidence of the very high price of errors and shortcomings when ensuring a nuclear power plant's safety. It means that there is a definite need for strict compliance with international safety-related requirements during the design, construction, and operation of nuclear power plants.

Additional information on the lessons learned from the Chernobyl accident is available in Reference 19.

Finally, with regard to the Fukushima accident, some of the general lessons learned are as follows [20]:

- Improve the nuclear safety-related culture.
- Seek out and effectively act on new information concerning hazards.
- Strengthen capabilities effectively to assess risk from beyond-design-basis events.
- Examine offsite emergency response-related capabilities and carry out appropriate improvements.
- Improve nuclear plant systems, resources, and training for enabling effective ad hoc responses in the event of severe accidents.
- Further incorporate up-to-date risk-related concepts into nuclear safety-related regulations.

Additional information on the lessons learned from the Fukushima accident is available in Reference 20.

7.8 Comparison of the Chernobyl and Three Mile Island Accidents and of the Fukushima and Chernobyl Accidents

The Chernobyl accident was due to willful violations of safe operating procedures and poor understanding of the effects of reactivity's positive void coefficient and minimum requirements of the control rod. In comparison, the Three Mile Island reactor No. 2 accident was basically the result of poor maintenance practice, poor operator training, and the lack of proper communication that could have alerted all operators involved with the reactor No. 2 regarding a quite similar incident at the Davis-Besse Nuclear Power Plant [21].

The Chernobyl accident led to a very large amount of radioactivity release to the surrounding environment with very serious health-related consequences, while the Three Mile Island accident had quite minimal actual radiological-related consequences, nevertheless causing quite serious psychological trauma in general to the public at large [21].

The comparison of Fukushima and Chernobyl accidents is as follows [22]:

- **Fukushima accident.** In this case, the power plant was not designed with consideration of the occurrence of a large tsunami. A major earthquake and tsunami together led to the destruction of all the power lines and backup generators. Once the power plant was without external power and all the generators were flooded, a very catastrophic decay heat casualty ensued, resulting in major reactor plant damage as well as meltdowns and explosive loss of containment of reactors.

- **Chernobyl accident.** In this case, the proximate cause was violation of involved procedures and human error. The unsafe reactor design resulted in instability at low power because of a positive void coefficient and formation of steam. When an improper test was performed around 1:00 AM at low power, the involved reactor became very critical. This was immediately followed by a steam explosion exposing the fuel, a core meltdown, as well as a raging graphite fire.

Additional information on the comparison of the Fukushima and Chernobyl accidents is available in Reference 22.

7.9 Problems

1. Write an essay on nuclear power plant accidents.
2. What are the levels of the INES? Describe each level in detail and who introduced the INES.
3. What were the world nuclear power plant accidents that have caused fatalities?
4. What were the INES rankings of the following nuclear power plant accidents:
    a. Chalk River accident
    b. Windscale Pile accident
    c. Tomsk accident
    d. Ishikawa accident
    e. Cadarache accident

5. Describe the nuclear power plant accident whose estimated cost was the highest.

6. What were the dates and the estimated costs of the following five accidents:
   a. Saclay accident
   b. Kalpakkam accident
   c. Plymouth accident
   d. Greifswald accident
   e. Vernon accident

7. Describe the Three Mile Island accident in detail.

8. Compare the Fukushima and Chernobyl accidents.

9. Discuss the lessons learned from the Three Mile Island and Chernobyl accidents.

10. Write an essay on the Fukushima accident.

References

1. Nuclear Reactor Accidents in the United States, retrieved on October 9, 2015 from website: https://en.wikipedia.org/wiki/Nuclear_reactor_accidents_in_the_United_States.

2. List of Nuclear Power accidents by Country, retrieved on August 15, 2016 from website: https://en.wikipedia.org/wiki/List_of_nuclear_power_accdents_by_country

3. Sovacool, B.K., A critical evaluation of nuclear power and renewable electricity in Asia, *Journal of Contemporary Asia*, Vol. 40, No. 3, 2010, pp. 393–400.

4. Sovacool, B.K., The costs of failure: A preliminary assessment of major energy accidents, 1907–2007, *Energy Policy*, Vol. 36, 2008, p. 1807.

5. Event Scale Revised for Further Clarity, retrieved on September 13, 2010 from website: http://www.World-nuclear-news.org. 6 October 2008

6. International Nuclear Event Scale, retrieved on August 15, 2016 from website: https://en.wikipedia.org/wiki/International_Nuclear_Event_Scale

7. Nuclear Power Plant Accidents: Listed and Ranked since 1952, retrieved on August 17, 2016 from website: https://www.thguardian.com/news/datablog/2011/mar/14/nuclear-power-plant-accidents-list-rank

8. Three Mile Island Accident, retrieved on August 19, 2016 from website: https://en.wikipedia.org/wiki/Three_Mile_Island_accident

9. Nuclear Regulatory Commission-Backgrounder on the Three Mile Island Accident, retrieved on August 19, 2016 from website: http://www.nrc.gov/reading-rm/doc-collections/fact-sheets/3mile-isle.html

10. Three Mile Island Accident, retrieved on August 20, 2016 from website: http://www.world-nuclear.org/informatin-library/safety-and-security/safety-of-plants/three-mile-island-accident.aspx

11. Executive Summary: Chernobyl: Assessment of Radiological and Health Impact, retrieved on August 23, 2016 from website: https://www.oecd-nea.org/rp/chernobyl/coe.html

12. Chernobyl Disaster, retrieved on August 19, 2016 from website: https://en.wikipedia,org/wiki/Chernobyl_disaster

13. Medvedev, Z.A., *The Legacy of Chernobyl*, W.W. Norton and Company, New York, 1990.

14. Lipscy, P., Kushida, K., Incerti, T., The Fukushima disaster and Japan's nuclear plant vulnerability in comparative perspective, *Environmental Science and Technology*, Vol. 47, 2013, pp. 6082–6088.

15. Fukushima Daiichi Nuclear Disaster, retrieved on August 19, 2016 from website: https://en.wikipedia.org/wiki/Fukushima_Daiichi_nuclear_disaster

16. Fackler, M., *Japan Power Company Admits Failings on Plant Precautions*, The New York Times, New York, October 12, 2012.

17. Fukushima Accident, retrieved on August 24, 2016 from website: http://www.world-nuclear.org/information-library/safety-and-security/safety-of-plants/fukushima-accident.aspx

18. Kessler, G. et al., *The Risks of Nuclear Technology*, Springer-Verlag, Berlin, 2014.

19. Lessons Learned from Chernobyl, retrieved on August 25, 2016 from website: http://www.tesec-int.org/chernobyl/Lessons.htm

20. Summary: Lessons Learned from the Fukushima Nuclear Accident for Improving Safety and Security of U.S. Nuclear Plants, retrieved on August 25, 2016 from website: http://www.ncbi.nlm.nih.gov/books/NBK253923/

21. Lee, J.C., McCormick, N.J., *Risk and Safety Analysis of Nuclear Systems*, John Wiley & Sons, Hoboken, New Jersey, 2011.

22. Comparison of Fukushima and Chernobyl Accidents, retrieved on August 23, 2016 from website: http://en.wikipedia.org/wiki/comparison_of_Fukushima_Chernobyl_nuclear_accidents

8

Reliability and Maintenance Programs for Nuclear Power Plants

8.1 Introduction

Over the years, the reliability and maintenance of nuclear power plants have become an important issue around the globe. Effective reliability and maintenance programs are essential for the safe operations of nuclear power plants.

The effective reliability program assures that the systems important to safety (SIS) shall satisfy their specified design and performance-related criteria at acceptable levels of reliability during the facility's lifetime [1]. The maintenance program of nuclear power plants consists of procedures, processes, and policies that provide appropriate directions to maintain nuclear power plant structures, systems, or components. Its range of maintenance-related activities includes the assessment, inspection, monitoring, testing, calibration, surveillance, overhaul, service, replacement, and repair of parts [2].

This chapter presents various important aspects of reliability and maintenance programs for nuclear power plants.

8.2 A Reliability Program's Objectives and Requirements

The objective of a reliability program is to ensure that all SIS at a nuclear power plant operate reliably, as per the relevant design and performance criteria, including any nuclear power plant's safety goals and regulatory body's license requirements.

With regard to the requirements, a reliability program for a nuclear power plant shall [1]:

1. Specify reliability-related targets for the systems considered important to safety at the nuclear power facility/plant.
2. Provide appropriate information to the maintenance program for maintaining the effectiveness of all SIS at the nuclear power plant/facility.

3. Include appropriate provisions for assuring, verifying, and demonstrating that the program is implemented properly.

4. Highlight and describe all the potential failure modes of all systems considered important to safety at the nuclear power plant/facility.

5. Provide for inspections, monitoring, tests, modeling or other actions for effectively assessing the reliability of all the systems considered important to safety at the nuclear power plant/facility.

6. Incorporate appropriate provisions to record and report the findings of program-related activities, including the findings of inspections, tests, assessments, or monitoring of the reliability of all the systems considered important to safety at the nuclear power plant/facility.

7. State the lowest performance and capability levels that the systems considered important to safety must attain for achieving reliabilities that are absolutely consistent with the safety targets of the nuclear power plant/facility and the requirements of the regulatory body.

8. Comprehensively and clearly record the reliability program's elements, attributes, activities, findings, and administration, including:

   a. The list of all systems considered important to safety at the nuclear power plant/facility.

   b. All the activities that make up the program.

   c. All the potential failure modes of the systems considered important to safety at the nuclear power plant/facility.

   d. The findings of the reliability assessments, monitoring, verifications, inspections, testing, and reporting-related activities that the licensee (i.e., the organization operating the nuclear power plant) performed as element of the reliability program.

   e. Schedules and procedures for performing activities of the program.

   f. Methods employed for determining the potential failure modes of the systems considered important to safety at the nuclear power plant/facility.

   g. The methodology employed for assigning, ranking, and identifying reliability targets to the systems considered important to safety at the nuclear power plant/facility.

   h. The reliability-related targets for each of the systems considered important to safety at the nuclear power plant/facility.

   i. The licensee's organization to implement and manage the program, including the involved staff members' specific positions, roles, and responsibilities.

   j. Reliability assessments, monitoring, testing, verifications, inspections, and recording and reporting activities that the licensee will perform for assuring, verifying, demonstrating, or documenting that the reliability program is implemented properly as per the regulatory body's requirements.

9. Highlight, utilizing a systemic approach, all systems considered important to safety by:

   a. Ranking all the highlighted structures, systems, and components with regard to their relative importance to safety.

   b. Screening out systems considered important to safety that do not contribute considerably to nuclear power plant safety.

   c. Highlighting nuclear power plant structures, systems, and components related to the detection, prevention, imitation, or mitigation of any failure sequence that could result in the damage of fuel, related release or radionuclide, or both.

8.3 Guidance for Developing Reliability Programs

The reliability program of a nuclear power plant for accomplishing its objective of enhancing plant availability and safety should possess the five elements shown in Figure 8.1 [1]. Furthermore, the reliability of the systems considered important to safety should be considered at power and shutdown states of the reactor and for all aspects of the reliability program, the impact of the postaccident mission time should also be considered.

The elements shown in Figure 8.1 can be accomplished through actions presented below [1].

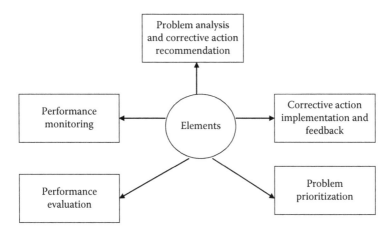

FIGURE 8.1
Elements for accomplishing the nuclear power plant reliability program's objective of enhancing plant availability and safety.

8.3.1 Using Systematic Approaches for Identifying and Ranking SIS

For identifying SIS, the probabilistic safety assessment (PSA) is the most thorough and systematic approach/method. The approach/method includes the insights from a Level-2 PSA, shutdown PSA, and external events and hazards-related assessments. However, it should be noted that when identifying SIS, other principles and information such as results of deterministic safety analysis, defense-in-depth, expert judgement, and operating experience should also be considered.

In the case of ranking SIS, the systems highlighted as important to safety should be ranked on the basis of their relative importance to safety as well as according to their contribution to the overall risk of the plant. This ranking should be carried out utilizing the findings of a plant-specific PSA.

The systems considered important to safety that do not contribute to the safety of the nuclear power plant may be screened out of the reliability program.

8.3.2 Specifying Reliability-Related Targets

The objective of setting these targets for systems considered important to safety is to develop a reference point against which to measure system performance. The reliability targets assigned to systems considered important to safety should be consistent with the safety goals of the nuclear power plant as well as they should factor in industry-wide operating experience where practicable.

A document entitled "Guide for General Principles of Reliability Analysis of Nuclear Power Generating Station Safety Systems" issued by the Institute of Electrical and Electronics Engineers (IEEE) provides the basis for establishing numerical reliability targets, that are based on the following three factors [1,3]:

- **Factor 1:** Risk
- **Factor 2:** Failure consequence
- **Factor 3:** Frequency of demand

Furthermore, a document entitled "Status, Experience and Future Prospects for the Development of Probabilistic Safety Criteria" issued by the International Atomic Energy Agency (IAEA), provides the principles for deriving numerical objectives [1,4]. The selection of reliability-related targets should maintain an effective balance between events' prevention and mitigation by applying the following two principles [1]:

1. The values of reliability targets for special safety systems should not be set lower than 0.999.
2. For all other poised systems considered important to safety, the value of the target should be set at or less than 120% of the system baseline performance.

8.3.3 Highlighting and Describing Potential Failure Modes

All the potential failure modes of systems considered important to safety should be clearly highlighted and described in order to determine necessary maintenance-related actions and ensure reliable operation of systems. Failure modes include failure to run for a stated mission time and failure to start on demand.

Failure modes can be highlighted through the following two ways [1]:

1. From failure history
2. Use of qualitative analytical methods when the failure history is not available

8.3.4 Stating Minimum Capabilities and Levels of Performance

For each and every success criterion of a system considered important to safety, the minimum performance levels and capabilities should be clearly specified and expressed in physical terms (e.g., voltage, intensity, flow, pressure).

Failure criteria of a system considered important to safety should be outlined in terms of the system not carrying out its operation when needed to do so. The failure criteria should be clearly consistent with the system failure criteria's definition utilized in all other analyses and/or other documentation that support the operating license. It should be noted that the system considered important to safety may have various different failure criteria, depending on the plant state, accident condition, or the failure consequences.

8.3.5 Maintenance Program

A maintenance program's basic objective is to maintain the nuclear power plant systems and equipment in accordance with applicable regulations, codes and standards, vendor recommendations, and previous experience so that their performance satisfies reliability targets effectively. The reliability modeling of systems considered important to safety provides useful information on how the maintenance program directly or indirectly affects the reliability of a system.

Consistent corrective maintenance and preventive maintenance may result in improvements in failure trends. Reliability-centered maintenance is one method that makes use of reliability principles for improving maintenance.

Information from the maintenance program is used in the modeling of the probability of failure of systems considered important to safety. Thus, the reliability modeling of systems considered important to safety provides quite useful information on how the maintenance program affects the reliability of a system.

8.3.6 Inspections, Tests, Modeling, and Monitoring

8.3.6.1 Providing for Inspections and Tests

Appropriate testing programs for systems considered important to safety should be developed and activated. In order to avoid introducing a common-cause failure where feasible, surveillance activities on redundant equipment should not be conducted at the same time or by the same personnel.

8.3.6.2 Modeling

The model used for describing the system should accurately reflect the current configuration of the system. The model's level of detail should be such that all dependencies are properly highlighted, but also limited to failure modes of equipment.

The information that should be included in the model is available in Reference 1.

8.3.6.3 Monitoring Performance and Reliability

As performance monitoring relies on collecting pertinent failure detection and in-plant reliability information, it includes both reliability monitoring (e.g., outage rate and observation of failure frequency) and condition monitoring (e.g., observation of conditions related to failure).

Thus, monitoring performance and reliability involves monitoring the performance of systems, monitoring the performance of components, monitoring human performance, and performing reliability assessments. Additional information on these four items is available in Reference 1.

8.3.7 Implementing a Reliability Program

In this case, the licensee (i.e., the operating organization of the nuclear power plant) should demonstrate its reliability program's effective implementation.

8.3.8 Recording and Reporting Results of Reliability Program-Related Activities

Results could be documented in the form of test results, work orders, operational logs, work plans, calibration records, and work permits. The review of this documented information is absolutely necessary for assuring accurate, timely assessment, and reporting of the reliability performance of systems considered important to safety.

8.3.9 Documenting a Reliability Program

For this no specific guidance is required.

8.4 A Maintenance Program's Objectives, Scope, and Background

As effective maintenance is important for the safe operation of a nuclear power plant, the objective of the maintenance program is to ensure that the regulatory body's maintenance-related requirements concerning nuclear power plant, by the licensee (i.e., operating organization of the nuclear power plant) are fully satisfied [2]. The range of activities in the maintenance program includes testing, monitoring, overhaul, calibration, surveillance, repair and replacement of parts, service, assessment, and inspection [3].

The maintenance program's scope covers each and every structure, system, or component within bounds of the nuclear power plant [2].

In order to ensure that structures, systems, or components function as per design, the nuclear power facility must be inspected, monitored, tested, assessed, and maintained. The most of maintenance-related activities are generally allocated to the preventive maintenance concept. These activities can be derived, for example, from the safety analysis-related assumptions, operating experience, codes and standards, and design or reliability requirements and are carried out on the basis of service time, predicted condition, or actual condition [2].

In situations in which the performance or condition of a structure, system, or component does not permit it to operate as per design, appropriate corrective action must be taken. The findings of all maintenance-related activities are fed back through an optimization process which enables the maintenance program's continuous improvement.

8.5 Guidance for Developing Maintenance Programs

In order to ensure that the nuclear power plant's overall maintenance strategy is effective, the maintenance program must include mutually supporting elements. The topics covered by these elements are shown in Figure 8.2 [2,3].

Each element of the program needs sufficient amount of resources governed by the approved policies, procedures, and processes of the licensee (i.e., the operating organization of the nuclear power plant). All these elements, when integrated, will form a quite comprehensive program. The elements' topics shown in Figure 8.2 are described below separately [2,3].

8.5.1 Program Basis

A systematic approach shall be followed for identifying which maintenance-related activities are to be carried out; on which structures, systems, or components; and at what intervals. The selection, frequency, and

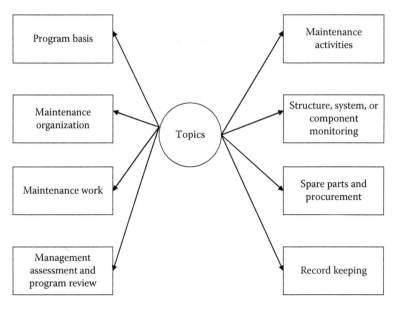

FIGURE 8.2
Topics covered by the maintenance program elements.

identification of maintenance-related activities shall take the following items into account [2,3]:

- All applicable industry code and standards requirements
- Reliability program-related requirements [1]
- The technical basis to demonstrate that safety-related goals and performance criteria are satisfied as presented in the licence and supporting documentation
- Design and operating conditions
- Operating experience
- The relative structure, system, or component importance of the risks to national security, the environment, and the health and safety of persons
- Aging management-related requirements
- Vendor recommendations
- The radiation protection principle as low as reasonably achievable (ALARA)

The objective of the maintenance program is to ensure that structures, systems, or components operate according to design, and it therefore follows that a maintenance-related strategy must have its basis on the approved plant design and safety-related analysis. Therefore, strategy development needs a very close liaison between the operating and design organizations for ensuring that the strategy is properly based on a clear comprehension of design philosophy and plant details.

The findings of maintenance-related activities shall be utilized to provide essential feedback for design modifications/program changes. Additional information on program basis is available in Reference 2.

8.5.2 Maintenance Organization

The operating organization of the nuclear power plant shall establish a maintenance organization for effectively implementing the maintenance program by considering the following five factors [2,3]:

- **Factor 1: Policies, Processes, and Procedures.** In this case, the operating organization of the nuclear power plant shall:
  - Set out policies, processes, and procedures that govern how the maintenance program is to be implemented.
  - Ensure that the maintenance policies, processes, and procedures are controlled, adhered to, and revised as required for reflecting the current plant configuration.
  - Have a process to ensure that all program documents are up to date.
- **Factor 2: Organizational Structure.** In this case, the senior management of the power plant shall be totally responsible for establishing and implementing the maintenance program. Furthermore, the senior management personnel shall establish clear lines of authority and define all the involved managerial and supervisory positions' responsibilities.

  Functions required for meeting program element requirements may not traditionally be carried out by groups within the maintenance department. In such situations, the interfaces between such groups shall be clearly highlighted.
- **Factor 3: Maintenance Facilities.** In this case, appropriate maintenance facilities and work areas shall be provided including appropriate housing for the receiving, shipping, handling, and storing of spare parts, equipment, and tools.
- **Factor 4: Training and Qualification of Workers.** In this case, the maintenance program shall be clearly supported by adequate numbers of fully trained and qualified personnel. The adequacy of such resources shall be determined on the basis of the maintenance program's objective.

  Training and qualification of all involved personnel shall be kept up-to-date. All involved maintenance personnel shall be provided facility-specific training in plant systems, work control, safety-related rules, radiation protection, access control, security, and emergency-related procedures that are commensurate with their assigned responsibilities.

- **Factor 5: Contract Workers.** In this case, the operating organization of the nuclear power plant shall ensure that all involved contractors clearly comply with work-related procedures and standards that are higher or equal to those applicable to plant staff personnel, particularly in the areas of professional competence, adherence to procedures, and performance evaluation.

   Furthermore, appropriate steps shall be taken for ensuring that all involved contract workers clearly conform to the operating organization's equivalent technical standards.

8.5.3 Maintenance Activities

The operating organization of the nuclear power plant shall include activities clearly aimed at detecting, avoiding, and repairing structure, system, or component failures. These activities are as follows [2,3]:

- **Preventive Maintenance.** A preventive maintenance program shall be established that clearly follows accepted industry practices and standards. Preventive maintenance-related activities are grouped as planned, predictive or periodic (time-based). Time-based preventive maintenance should not be scheduled just before functional/performance testing as this may mask equipment/system degradation.

- **Corrective Maintenance.** The operating organization of the power plant shall have appropriate processes in place to initiate corrective maintenance and conduct equipment failure diagnosis. These processes shall include the failed equipment's impact as well as prioritizing the repair work with regard to all ongoing maintenance-related activities.

   The operating organization of the power plant shall have an appropriate process in place to control and perform temporary repairs including appropriate approvals, equivalency assessments, and time period until the permanent repair can be executed or an approved modification carried out.

- **Activity Optimization.** A process shall be in place for optimizing the maintenance activities on the basis of results from items such as failures in operation, failures found during maintenance activities, operating experience in other plants, the as-found condition, maintainability improvements, and frequency of faults.

- **Aging Management.** The operating organization of the power plant shall have a process for assessing, detecting, and managing structure, system, or component deterioration as a result of aging effects such as erosion, irradiation, fatigue, corrosion, and other material-related degradation. The frequency and types of maintenance-related activities shall be changed for accommodating such effects.

8.5.4 Structure, System, or Component Monitoring

The operating organization of the nuclear power plant shall establish appropriate baseline criteria against which structure, system, or component performance and function can be measured. The criteria should include availability, reliability, performance, and function requirements used in safety analysis and the plant design. Furthermore, the degree of structure, system, or component monitoring shall commensurate with the structure's, system's, or component's safety significance.

Condition monitoring, testing, and surveillance concerning structure, system, or component monitoring are described below separately [2,3].

- **Condition Monitoring.** The operating organization of the power plant shall have appropriate procedures and processes for performing condition monitoring. These include items such as periodic and in-service inspections, measurements or trending of structure, system or component performance or physical characteristics for indicating current condition as well as future potential for failure.

  Generally, condition monitoring is performed on a nonintrusive basis and includes the employment of specialized equipment. Two examples of condition monitoring techniques are vibration analysis and thermography.

- **Testing.** The operating organization of the power plant shall have appropriate procedures and processes for performance and function testing for verifying that structures, systems, or components are in proper working order as well as in a state of readiness for carrying out their functions. The operating organization should develop appropriate test plans, and such plans should incorporate the test frequency and acceptance criteria.

  Instrumentation and test equipment used in conducting the test program shall have the range and accuracy required for demonstrating that acceptance criteria have been fully satisfied.

- **Surveillance.** The operating organization of the power plant shall have appropriate procedures and processes to conduct structure, system, or component surveillance and the surveillance's results shall be documented appropriately. Some examples of surveillance are process system configuration checks, routine readings, and system walkdowns.

Additional information on condition monitoring, testing, and surveillance is available in Reference 2.

8.5.5 Maintenance Work

The operating organization of the nuclear power plant shall have appropriate procedures and processes to initiate, assess, manage, prioritize, plan, and schedule maintenance work. The resulting maintenance activity-related schedules shall be reviewed regularly and revised appropriately to account for changing conditions, operating experiences, and modifications.

The organization shall properly undertake and accomplish all maintenance-related activities in such a way that is commensurate with the structure's, system's, or component's safety significance as well as with effective allocation or resources.

The identification of unit and equipment shall be clear in all work procedures as well as in the field for ensuring that the proper equipment/system is isolated, maintained, and returned to service. In addition, an effective process for foreign material exclusion shall be in place as appropriate for all maintenance-related work.

Work assessment; maintenance procedures; work planning, scheduling. and execution; outage activities; and postmaintenance verification and testing are described below separately [2,3].

- **Work Assessment.** An appropriate process shall be implemented to assess maintenance activities. The process, in addition to job tasks, shall assess the maintenance activities' impact on safety including items such as regulatory requirements, safe operating envelope, and all potential industrial and radiological hazards to site manpower, the environment, and public-at-large.

- **Maintenance Procedures.** Maintenance shall be carried out in accordance with clearly approved written procedures, instructions, or drawings as appropriate to the given situation. Thus, an appropriate process should be in place to control procedure preparation, review, validation, issue, modification, and revision.

   In situations, when a procedure for performing a maintenance task is found to be inadequate, there shall be appropriate actions for ensuring that the task is halted or safely managed until such time as the inadequacy in the procedure is corrected properly.

- **Work Planning, Scheduling, and Execution.** Appropriate procedures and processes shall be implemented to plan, schedule, and execute all maintenance activities. Work planning should take place at the individual job level as well as at the overall plant level.

   Procedures for dealing with any scheduled maintenance activity's omission or deferral shall be included. All maintenance work-related activities should be carried out as per the approved work package.

- **Outage Activities.** An appropriate process shall be implemented for managing the increased maintenance-related activities during plant outages. During an outage, the plan for taking equipment out of service for maintenance shall include appropriate measures for dealing with an event's all possible consequences occurring while the equipment is still out of service.

- **Postmaintenance Verification and Testing.** Prior to returning equipment to an operational state, the operating organization of the nuclear power plant shall properly ensure that postmaintenance verification has been fully accomplished, the affected configuration is verified, all relevant records are properly reviewed for completeness and any unexpected discoveries have been appropriately assessed and dispositioned.

Additional information on maintenance work is available in Reference 2.

8.5.6 Spare Parts and Procurement

The operating organization of the nuclear power plant shall establish appropriate procedures and processes for procuring, receiving, storing, securing, and issuing spare parts, tools, and materials. The spares shall fully meet the same technical standards and quality-related requirements as the already installed items in the power plant.

All procurement processes shall clearly include requirements for all qualified suppliers. The acceptance procedures and receipt shall incorporate a requirement to label, tag, and quarantine noon-conforming items.

8.5.7 Management Assessment and Program Review

The operating organization of the nuclear power plant shall develop a continuous process for assessing, reviewing, and improving the maintenance program to ensure that the maintenance strategy is effective, fully satisfies its objectives, and has been implemented as per the all applicable industry standards and codes. Whenever a maintenance program-related deficiencies is highlighted, its significance shall be properly assessed and where appropriate a cause determination shall be carried out and appropriate corrective measures taken.

Each and every assessment and review shall be recorded and documented properly.

Additional information on management assessment and program review is available in Reference 2.

8.5.8 Record Keeping

Record keeping is very important in order to meet a regulatory body's requirements such as specified in Reference 4. Thus, records and reports shall include sufficient information for providing objective evidence that the maintenance program is being totally implemented as well as in accordance with the specified quality assurance program.

The operating organization of the nuclear power plant shall properly document a description of all repairs performed, highlighting the component/part that malfunctioned, the failure cause, the remedial measure taken, and the system's state after repairs.

Additional information on record keeping is available in Reference 2.

8.6 Reliability and Maintenance Program-Related Standards

There are a large number of national and international standards, directly or indirectly, concerned with reliability and maintenance programs at nuclear power plants. Some of these standards are as follows [1,2,5]:

- Guide for the Definition of Reliability Program Plans for Nuclear Power Generating Stations, and other Nuclear Facilities, IEEE 933-2013, Institute of Electrical and Electronics Engineers (IEEE) Standards Association, Piscataway, New Jersey, United States, 1999.
- Guide for General Principles of Reliability Analysis of Nuclear Power Generating Station Safety Systems, IEEE 352-1987, IEEE Standards Association, Piscataway, New Jersey, United States, 1987.
- Safety Related Maintenance in the Framework of the Reliability Centered Maintenance Concept, IAEA TECDOC-658, International Atomic Energy Agency (IAEA), Vienna, Austria, 1997.
- Guidance for Optimizing Nuclear Plant Maintenance Programmes, IAEA TECDOC-1383, International Atomic Energy Agency, Vienna, Austria, 2003.
- Safety Culture in the Maintenance of Nuclear Power Plants, IAEA Safety Standards Series No. 42, International Atomic Energy Agency, Vienna, Austria, 2005.
- Good Practices for Cost Effective Maintenance of Nuclear Power Plants, IAEA TECDOC-928, International Atomic Energy Agency, Vienna, Austria, 1997.
- Maintenance, Surveillance and In-Service Inspection in Nuclear Power Plants, IAEA Safety Standards Series No. NS-G-2.6, International Atomic Energy Agency, Vienna, Austria, 2002.

- Application of the Single-Failure Criterion to Nuclear Power Generating Station Safety Systems, IEEE 379-2014, Institute of Electrical and Electronics Engineers Standards Association, Piscataway, New Jersey, USA, 2011.
- Standard Requirements for Reliability Analysis in the Design and Operation of Safety Systems for Nuclear Power Generating Stations, IEEE 577-2012, Institute of Electrical and Electronics Engineers Standards Association, Piscataway, New Jersey, USA, 2012.
- Guide for Incorporating Human Action Reliability Analysis for Nuclear Power Generating Stations, IEEE 1082-1997, Institute of Electrical and Electronics Engineers Standards Association, Piscataway, New Jersey, USA, 1997.
- Administrative Practices for Nuclear Criticality Safety, ANSI/ANS-8.19-1996, American Nuclear Society, La Grange Park, Illinois, USA, 1996.
- External Human Induced Events in Site Evaluation for Nuclear Power Plants, IAEA Series No. NS-G-3.1, International Atomic Energy Agency, Vienna, Austria, 2002.
- Commissioning for Nuclear Power Plants, IAEA Series No. SSG-28, International Atomic Energy Agency, Vienna, Austria, 2014.
- Decommissioning of Nuclear Power Plants and Research Reactors, IAEA Series No. WS-G-2.1, International Atomic Energy Agency, Vienna, Austria, 1999.
- Maintaining the Design Integrity of Nuclear Installations Throughout Their Operating Life, IAEA INSAG-19, International Atomic Energy Agency, Vienna, Austria, 2003.
- Maintenance of Nuclear Power Plants, IAEA Safety Standards Series No. 50-SG-07, International Atomic Energy Agency, Vienna, Austria, 1990.
- Regulatory Surveillance of Safety Related Maintenance at Nuclear Power Plants, IAEA TECDOC-960, International Atomic Energy Agency, Vienna, Austria, 1997.

8.7 Problems

1. What is the objective of the reliability program concerning a nuclear power plant?
2. List at least six requirements concerning a nuclear power plant's reliability program.

3. What are the main elements of the reliability program of a nuclear power plant for accomplishing its objective of enhancing plant availability and safety?

4. Discuss the actions for accomplishing the main elements of the reliability program of a nuclear power plant.

5. Discuss nuclear power plant maintenance programs' objective.

6. What are the topics covered by the nuclear power plant maintenance program elements?

7. Discuss preventive and corrective maintenance concerning nuclear power plants.

8. Compare reliability programs with maintenance programs concerning nuclear power plants.

9. List at least five IAEA standards/documents concerning nuclear power plant maintenance.

10. List at least five national/international standards/documents concerning nuclear power plant reliability.

References

1. Reliability Programs for Nuclear Power Plants, Report No. RD/GD-98, Canadian Nuclear Safety Commission, Ottawa, Canada, 2012.
2. Maintenance Programs for Nuclear Power Plants, Report No. RD/GD-210, Canadian Nuclear Safety Commission, Ottawa, Canada, 2012.
3. Maintenance Programs for Nuclear Power Plants, Report No. S-210, Canadian Nuclear Safety Commission, Ottawa, Canada, 2007.
4. Reporting Requirements for Operating Nuclear Power Plants, Report No. S-99, Canadian Nuclear Safety Commission, Ottawa, Canada, 2003.
5. Nuclear Power Standards, retrieved on October 5, 2016 from website: https://standards.ieee.org/findstds/standard/nuclear_power_all.html.

9

Human Factors and Human Error in Nuclear Power Generation

9.1 Introduction

Although human factors play an important role in nuclear power generation, it was not until the mid-1970s when nuclear power plants were designed by utilizing human factors design-related standards or analytical methods [1,2]. Prior to this, the nuclear power industrial sector utilized the same engineering methods that had been developed over the past many decades in designing hydro and fossil fuel power generation plants [1–4].

A study conducted by one utility clearly indicates that the occurrence of failures due to human error is approximately two and half times higher than those due to hardware failures [5]. As the humans play an important role during design, production, operation, and maintenance phases of systems used in nuclear power plants, the occurrence of human error in nuclear power generation can be quite catastrophic. An example is the Chernobyl Nuclear Power Plant accident in Ukraine that occurred on April 26, 1986, directly or indirectly due to human error [6].

This chapter presents various important aspects of human factors and human error in nuclear power generation.

9.2 Aging Nuclear Power Plant Human Factor-Related Issues

Humans play a very important role in the effects of aging on reliable and safe operation of nuclear power stations/plants. Thus, human factor-related issues may be more important than the issues concerning the degradation of components and materials with age as human actions can accelerate or decelerate a nuclear power plant's physical aging. These human factor-related issues can be grouped under six classifications [7]: maintenance, management, staffing, training, procedures, and man-machine interface.

Maintenance will impact and be impacted by aging nuclear power stations/plants. Some of the human factor-related issues concerning maintenance are that new maintenance staff personnel may not be properly trained in the technology that exists in aging power plants, increment in dose rates will make maintenance activities in certain areas of older power plants more difficult to perform, and loss in flexibility to transfer maintenance staff personnel to other power plants due to significant technological-related differences between new and old power plants [7].

Past experiences clearly indicate that management impacts basically all facets of power plant performance including items such as technical support, operations, construction and modifications, maintenance, and design [8]. In aging power plants, there are many management-related issues that may arise. These issues include lack of interest by top management, tendency to delay decisions as long as possible, and difficulties in retaining and attracting good managers [9].

In power plant aging, the staffing-related issues are concerned with the ability for attracting and retaining good personnel for older power plants, particularly in areas such as technical support and operation.

The remaining three classification issues (i.e., procedures, training, and man-machine interface) are considered quite self-explanatory; however, useful information on all these issues is available in Reference 7.

9.3 Human Factor-Related Issues That Can Have a Positive Impact on the Decommissioning of Nuclear Power Plants

There are hundreds of nuclear reactors operating around the globe, and many of them are in the process of decommissioning [10,11]. Four human factor-related issues that can have the most positive impact upon safe and successful decommissioning of nuclear power plants are as follows [2,11]:

- Maintaining safety culture
- Uncertainty about the future
- Retaining organizational memory
- Maintaining adequate competence for decommissioning

Each of the above issues is described below, separately.

9.3.1 Maintaining a Safety Culture

Safety culture is fundamental to the nuclear industry, and the importance of its positive maintenance is an established element of daily regimen of all nuclear power stations/plants. The loss of key staff personnel and uncertainty can directly or indirectly have a profound impact on the safety culture of staff personnel left behind. As existing safety-related procedures, processes, and defenses may no longer be effective, the following actions could be very helpful for maintaining a positive safety culture during decommissioning of nuclear power plants [2,11–14]:

- Provide appropriate feedback on performance to all involved personnel on a regular basis to encourage motivation and morale as well as to highlight and raise awareness of all possible risks.
- Reevaluate all safety culture-related indicators throughout the decommissioning process for reflecting changing and new activities, environments, and risks as well as for ensuring accurate measurement of safety culture.
- Present decommissioning to all concerned personnel as a very good opportunity for their future career development rather than a threat for their future employment.
- Emphasize the importance of all involved personnel's contribution with regard to safe and efficient decommissioning.

9.3.2 Uncertainty about the Future

Uncertainty about the future can directly or indirectly affect decommissioning because of the subsequent impact it can have on morale and motivation of staff members. When the decision is made to shut down and decommission a nuclear power plant, uncertainty to a certain degree is to be expected because to most plant staff personnel, decommissioning will be a totally new phenomenon [11]. There are many different ways in which the uncertainty can manifest itself including reducing the involved staff personnel motivation, morale, and commitment to the power plant.

The most effective approach for managing uncertainty is lowering it to as low as possible [15]. This can be achieved by developing an effective communication strategy whereby regular and accurate information is provided to all involved staff personnel concerning future plans for decommissioning and the power plant.

Additional information on uncertainty about the future is available in References 2 and 13.

9.3.3 Retaining Organizational Memory

Retaining organizational memory requires clear and accurate knowledge of the power plant history and operations including any changes to the plant design. Most of this type of information can be obtained from the power plant-related documentation such as event reports, drawings, and procedures. Thus, it is essential to conduct a documentation audit as soon as possible for finding out what documentation is available as well as the accuracy of the information contained in the documentation. This information should be catalogued and stored in such a way so that it is easily accessible to all involved staff members.

The knowledge held by experienced staff members of the power plant may not have been documented properly through any formal processes or procedures. Because of uncertainty regarding the future, there is a considerable risk of losing these staff members in the early stages of planning for decommissioning. Therefore, it is important, as soon as possible prior to the power plant closure, to highlight key roles for decommissioning as well as to develop appropriate strategies for retaining the key staff members for these roles.

Additional information on retaining organizational memory is available in References 2 and 11.

9.3.4 Maintaining Adequate Competence for Decommissioning

Maintaining adequate competence for decommissioning can have the most impact upon successful and safe decommissioning and is composed of three stages. These three stages are as follows [11]:

- **Stage A:** This is concerned with initial decommissioning and includes items such as defueling and removal of all nonfixed contaminated parts/components and readily removable systems/equipment.
- **Stage B:** This is concerned with decontamination and dismantling of contaminated and other internal equipment/systems.
- **Stage C:** This is concerned with demolition of buildings and structures that are no longer needed.

Each of the above three stages will require different skills and knowledge of the nuclear power plant as well as of the technologies and tools/methods to be employed and thus a completely different workforce profile. It means that during the decommissioning process, a significant training program will need to be developed for satisfying the training needs of all involved staff personnel, contractors, and managers. Therefore, it is absolutely essential that appropriate training-related strategies are developed right at the planning stage so that the required number of trainers are retrained for developing and delivering training-related plans during the power plant's decommissioning process.

9.4 Human Factors Engineering Design Goals with Regard to Nuclear Power Generation Systems and Responsibilities

In regard to nuclear power generation systems, there are many human factors engineering design goals. Some of these goals are as follows [2,16]:

- All personnel-related tasks can be carried out within the stated time and performance criteria.
- The human-system interfaces, training, procedures, management and organizational support, and staffing/qualifications will effectively support a high degree of situation awareness.
- The plant design and allocation of functions will effectively support operation vigilance as well as will provide acceptable levels of workload (i.e., to minimize operator's overload and underload periods).
- All operator-related interfaces will minimize operator interfaces will minimize operator errors and will effectively provide for error detection and recovery capability.

There are many responsibilities of staff members who are responsible for human factors engineering within a nuclear power generation organization. Some of the important responsibilities are as follows [2,16]:

- Assuring that all human factors engineering-related activities clearly comply with the human factors engineering-related plans and procedures.
- Developing appropriate human factors engineering-related plans and procedures.
- Recommending, initiating, and providing appropriate solutions through designated channels for difficulties/problems highlighted during the human factors engineering-related activities' implementation.
- Verifying the team recommendations' implementation.
- Scheduling human factors engineering-related activities and milestones with regard to other modification-associated activities.
- Oversighting and reviewing human factors engineering-related design, development, test, and evaluation activities.

9.5 Human Factors Review Guide for Next-Generation Nuclear Reactors

In order to serve as a human factors review guide for the next generation of reactors, in 1994, the U.S. Nuclear Regulatory Commission (NRC) developed a document (NUREG-0711) entitled the "Human Factors Engineering Program Review Model" [17]. The document describes objective, background, applicant submittals, and review criteria for the following 10 elements [17,18]:

- **Element 1:** Human-System Interface Design
- **Element 2:** Human Factors Engineering Program Management
- **Element 3:** Functional Requirement Analysis and Function Allocation
- **Element 4:** Procedure Development
- **Element 5:** Operating Experience Review
- **Element 6:** Task Analysis
- **Element 7:** Human Reliability Analysis
- **Element 8:** Staffing
- **Element 9:** Training Program Development
- **Element 10:** Human Factors Verification and Validation

For each of the above 10 elements, the inputs considered useful are presented below, separately [16,17].

Element 1: Human-System Interface Design
In this case, the following inputs are considered useful:

- Ensure that the results of task analysis are properly reflected in all human-system interface designs.
- Conduct human-system interface evaluation independent of the human factors validation and verification.

Element 2: Human Factors Engineering Program Management
In this case, the following inputs are considered useful:

- In the human factors engineering program as human factors engineering issue tracking is a very important task; issue selection criteria, issue analysis methods, and issue handling and management procedures must be described in considerable detail.
- Human factors engineering–associated tasks, organization as well as relationships among these tasks should be reviewed with care.
- All applicants should submit the appropriate human factors engineering program plan for review by the regulatory agencies/bodies.

Element 3: Functional Requirement Analysis and Function Allocation
In this case, the following inputs are considered useful:

- Ensure that functional requirement analysis is conducted to the level with which function allocation can be carried out properly.

- In addition to highlighting functions important to safety, other general functions should also be included in the analysis for reflecting the findings of functional requirement analysis to the design.

Element 4: Procedure Development
In this case, the following inputs are considered useful:

- Ensure that appropriate guidelines for management, validation, and modification of procedures are developed when using computerized procedures.
- Ensure that interface with other information-related controls and displays is quite efficient for enhancing operator performance and not to induce human errors.

Element 5: Operating Experience Review
In this case, the following inputs are considered useful:

- As near-miss cases can provide information as important as that of plant-related events, it is considered very important to include the near-miss case analysis into the operating experience review.
- All important issues in the operation of earlier nuclear power plants must be analyzed carefully, and the findings must be reflected in the design under consideration.

Element 6: Task Analysis
In this case, the following inputs are considered useful:

- During the task analysis process, consider seriously the tasks expected to go through changes.
- Ensure that documentation clearly shows relationships between task analysis and functional requirement-related analysis and function allocation end results

Element 7: Human Reliability Analysis
In this case, the following inputs are considered useful:

- Ensure that the design-related data from probabilistic risk analysis required by the human factors engineering design team members are appropriately highlighted and all types of interactions between probabilistic risk analysis team members and the human factors engineering design team members are properly described.
- Ensure that a definition of data input-out between human factors engineering design team members and probabilistic risk analysis team members is clearly described.

Element 8: Staffing
In this case, the following input is considered useful:

- Because of the changes in human-system interface design and resulting personnel-related tasks, ensure that staffing analysis is conducted with utmost care.

Element 9: Training Program Development
In this case, the following input is considered useful:

- Ensure that all types of training-related requirements are clearly highlighted during the design phase.

Element 10: Human Factors Verification and Validation
In this case, the following input is considered useful:

- For this element, use the U.S. NRC document entitled "Integrated System Validation: Methodology and Review Criteria" [19].

9.6 Human Error Facts, Figures, and Examples Concerning Nuclear Power Generation

Some of the human error facts, figures, and examples directly or indirectly concerning nuclear power generation are as follows:

- A study of Licensee Event Reports performed by the U.S. NRC reported that approximately 65% of nuclear system failures were due to human error [20,21].
- As per References 22 and 23, approximately 70% of nuclear plant operation-related errors appear to have a human factors origin.
- As per References 24 and 25, a study of 143 occurrences of operating U.S. commercial nuclear power plants from February 1975 to April 1975 revealed that approximately 20% of the occurrences were due to operator-related errors.
- As per Reference 5, approximately 27% of the commercial nuclear power plant outages in the United States during the period from 1990–1994 were the result of human error.
- As per Reference 6, in 1986, the Chernobyl Nuclear Power Plant in Ukraine, widely referred to as the worst accident in the history of nuclear power, was due to human-related problems.

- In Japan, during the period from 1969–1986, around 54% of the incidents due to human errors resulted in automatic shutdown of nuclear reactors, and 15% of them led to power reduction [26].
- The Three Mile Island Nuclear Power Plant accident that occurred in 1979 in the United States was the result of human-related problems [6].
- A study of the major incident/accident reports of nuclear power plants in South Korea revealed that around 20% of the events occur due to human error [27].
- As per References 28 and 29, during the period from 1978–1992, of the 255 shutdowns that occurred in South Korean nuclear power plants, 77 of them were human-induced.

9.7 Occurrences Caused by Operator Errors during Operation in Commercial Nuclear Power Plants

Over the years, in commercial nuclear power plants during operation, there have been many occurrences due to operator errors. Twelve of these occurrences are as follows [24,25,30]:

- **Occurrence 1:** Sudden release of low-level radioactive water
- **Occurrence 2:** Reactor mode switch in wrong mode
- **Occurrence 3:** Two adjacent control blades withdrawn during the rod driver overhaul process
- **Occurrence 4:** Control rod not declared inoperable or nonfunctional when misaligned
- **Occurrence 5:** Reactor coolant system leakage evaluation not carried out
- **Occurrence 6:** Suppression chamber water volume goes beyond the limit
- **Occurrence 7:** An inadvertent power board trip
- **Occurrence 8:** Concentrated boric acid storage tank below limits
- **Occurrence 9:** Process vent gaseous radiation monitor left in calibrate mode
- **Occurrence 10:** Steam generator blow-down function not monitored
- **Occurrence 11:** Serious judgement-related error in monitoring
- **Occurrence 12:** Control rod inserted well beyond limit

Additional information on the above occurrences is available in Reference 24.

9.8 Causes of Operator Errors in Commercial Nuclear Power Plant Operations

Some of the causes for the occurrence of operator errors in commercial nuclear power plant operations are as follows [2,24]:

- Misunderstanding of procedures
- Use of incorrect procedures
- Disregard of procedures
- Procedural deficiency
- Misidentification of an alarm
- Operator oversight
- Lack of proper guidelines
- Inadequate procedures and lack of clarity
- Misunderstanding of technical specification
- Incomplete check list
- Typographical error

Additional information on the above causes is available in Reference 24.

9.9 Problems

1. Write an essay on human factors and human error in nuclear power generation.
2. What are the classifications of aging nuclear power plant human factor-related issues? Discuss each of these classifications.
3. What are the human factor-related issues that can have a positive impact on the decommissioning of nuclear power plants?
4. What are the human factors engineering design goals with regard to nuclear power generation systems?
5. What are the responsibilities of staff members who are responsible for human factors engineering within a nuclear power generation organization?
6. Describe the document (NUREG-0711) entitled the "Human Factors Engineering Program Review Model" developed by the U.S. NRC in 1994.

7. Describe at least six human error facts, figures, and examples that directly or indirectly concern nuclear power generation.

8. List at least eight occurrences caused by operator errors during operation in commercial nuclear power plants.

9. List at least eight causes of operator errors in commercial nuclear power plant operations.

10. Describe at least one issue in detail that can have a positive impact on the decommissioning of nuclear power plants.

References

1. Parsons, S.O., Seminara, J.L., Power systems human factors/ergonomics activities in the United States, *Proceedings of the IEA/HFEA Congress*, 2000, pp. 3.807–3.810.

2. Dhillon, B.S., *Human Reliability, Error, and Human Factors in Power Generation*, Springer, London, 2014.

3. Seminara, J.L., Parsons, S.O., Survey of control-room design practices with respect to human factors engineering, *Nuclear Safety*, Vol. 21, No. 5, 1980, pp. 603–617.

4. Parson, S.O., Eckert, S.K., Seminara, J.L., Human factors design practices, for nuclear power plant control rooms, *Proceedings of the Human Factors Society Annual Meeting*, Detroit, Michigan, 1978, pp. 133–139.

5. Varma, V., Maintenance training reduces human errors, *Power Engineering*, Vol. 98, 1996, pp. 44–47.

6. Kawano, R., Steps toward the realization of human-centered systems: An overview of the human activities at TEPCO, *Proceeding of the IEEE Sixth Annual Human Factors Meeting*, 1997, pp. 13.27–13.32.

7. Widrig, R.D., Human factors: A major issue in plant aging, *Proceedings of the International Conference on Nuclear Power Plant Aging*, 1985, pp. 65–68.

8. Osborn, R., et al., Organizational Analysis and Safety for Utilities with Nuclear Power Plants, Report No. NUREG/CR-3215, United States Nuclear Regulatory Commission, Washington, DC, 1983.

9. Child, J., Kieser, A., Development of organizations overtime, in *Handbook of Organizational Design*, Oxford University Press, New York, 1981, pp. 28–64.

10. *Nuclear Power Reactors in the World*, IAEA 2007 Edition, International Atomic Energy Agency (IAEA), Vienna, Austria, June 2007.

11. Blackett, C., The role of human factors in planning for nuclear power plant decommissioning, *Proceedings of the 3rd IET International Conference on System Safety*, 2008, pp. 1–6.

12. Planning, Managing, and Organizing the Decommissioning of Nuclear Facilities: Lessons Learned, Report No. IAEA-TECDOC-1394, International Atomic Energy Agency, Vienna, Austria, May 2004.

13. Managing Change in Nuclear Facilities, Report No. IAEA-TECDOC-1226, International Atomic Energy Agency, Vienna, Austria, July 2001.

14. Managing the Socio Economic Impact of the Decommissioning of Nuclear Facilities, IAEA Technical Report Series No. 464, International Atomic Energy Agency, Vienna, Austria, April 2008.

15. Durbin, N.E., Lekberg, A., Melber, B.D., Potential of Organizational Uncertainty on Safety, SKI Report No. 01:42, Swedish Nuclear Power Inspectorate, Stockholm, Sweden, December 2001.

16. Hill, D., Alvarado, R., Heinz, M., Fink, B., Naser, J., Integration of human factors engineering into the design change process, *Proceedings of the American Nuclear Society International Topical Meeting on Nuclear Pant Instrumentation, Controls, and Human-Machine Interface Technologies*, 2004, pp. 1–10.

17. Human Factors Engineering Program Review Model, Report No. NUREG-0711, U.S. Nuclear Regulatory Commission, Washington, DC, 1994.

18. Lee, J.W. et al., Human factors review guide for Korean next generation reactor, *Transactions of the 15th Conference on Structural Mechanics in Reactor Technology*, 1999, pp. XI-353–XI-360.

19. O'Hara, J., Stubler, J., Higgins, J., Brown, W., *Integrated System Validation: Methodology and Review Criteria*, Report No. NUREG/CR-6393, U.S. Nuclear Regulatory Commission, Washington, DC, 1997.

20. Trager, T.A., Jr., Case Study Report on Loss of Safety System Function Events, Report No. AEOD/C504, United States Nuclear Regulatory Commission, Washington, DC, 1985.

21. Ryan, G.T., A task analysis-linked approach of integrating the human factor in reliability assessments of nuclear power plants, *Reliability Engineering and System Safety*, Vol. 22, 1988, pp. 219–234.

22. An Analysis of Root Causes in 1983 and 1984 Significant Event Reports, Report No. 85–027, Institute of Nuclear Power Operations, Atlanta, Georgia, July 1985.

23. Williams, J.C., A data-based method for assessing and reducing human error to improve operational performance, *Proceedings of the IEEE Conference on Human Factors and Power Plants*, 1988, pp. 436–450.

24. Husseiny, A.A., Subry, Z.A., Analysis of human factor in operation of nuclear power plants, *Atomkernenergie Kerntechnik*, Vol. 36, 1980, pp. 115–121.

25. Scott, R.L., Recent occurrences of nuclear reactors and their causes, *Nuclear Safety*, Vol. 16, 1975, pp. 496–497.

26. Mishima, S., Human factors research program: Long term plan in cooperation with Government and Private Research Centers, *Proceedings of the IEEE Conference on Human Factors and Power Plants*, 1988, pp. 50–54.

27. Kim, J., Park, J., Jung, W., Kim, J.T., Characteristics of test and maintenance human errors leading to unplanned reactor trips in nuclear power plants, *Nuclear Engineering Design*, Vol. 239, 2009, pp. 2530–2536.

28. Lee, J.W., Park, G.O., Park, J.C., Sim, B.S., Analysis of human error trip cases in Korean NPPs, *Journal of Korean Nuclear Society*, Vol. 28, 1996, pp. 563–575.

29. Heo, G., Park, J., A framework for evaluating the effects of maintenance-related human errors in nuclear power plants, *Reliability Engineering and System Safety*, Vol. 95, 2010, pp. 797–805.

30. Thompson, D., Summary of recent abnormal occurrences at power-reactor facilities, *Nuclear Safety*, Vol. 15, 1974, pp. 198–220.

10

Human Factors and Human Error in Nuclear Power Plant Maintenance

10.1 Introduction

In nuclear power plant maintenance, human factors play an important role because improving the maintainability design of power plant facilities, equipment, and systems with regard to human factors helps to directly or indirectly increase plant availability, safety, and productivity. Interest in human factor-related issues in the nuclear power industrial sector is relatively new in comparison to the aerospace industrial sector. Its history goes back to the 1970s when the WASH-1400 Reactor Safety Study criticized the deviation of controls' and displays' design in the commercial nuclear power plants of the United States from the human factors engineering-related set standards [1–3].

A number of studies performed over the years clearly indicate that the occurrence of human error in maintenance is an important factor in nuclear power plant safety-related incidents [4,5]. For example, a study concerning nuclear power plant operating experiences reported that because of human errors in the maintenance of some motors in the rod drives, a number of motors ran backward and withdrew rods instead of inserting them [5]. Needless to say, over the years, the occurrence of many human factors' shortcomings-related events and human errors in nuclear power plants has led to an increased attention to human factors and human errors in nuclear power plant maintenance.

This chapter presents various important aspects of human factors and human errors in the area of nuclear power plant maintenance.

10.2 Study of Human Factors in Power Plant Maintenance

A study concerning the maintenance of five nuclear and four fossil-fuel power plants with respect to human factors reported various types of directly or indirectly human factor-related problems [6,7]. The study was quite wide-ranging in scope, extending to an examination of items such as environmental factors, organizational factors, designs, tools, spares, facilities, and procedures.

The findings of the study were grouped under equipment maintainability; maintenance, information, procedures, and manuals; facility design factors; anthropometrics and human strength; environmental factors; communications; personal safety; movement of humans and machines; labeling and coding; maintenance stores, supplies, and tools; maintenance errors and accidents; preventive maintenance and malfunction diagnosis; equipment protection; productivity and organizational interfaces; job practices; and selection and training. The first seven of these groups are described below [6,7].

The equipment maintainability's most common problem is the placement of equipment parts in locations that are inaccessible from a normal work position. The maintenance information, procedures, and manuals' main problem is inadequate manuals and poorly written procedures. The facility design factor problems include poor temperature–ventilation control, high noise levels, insufficient storage space to satisfy maintenance needs effectively, and inadequate facilities to store contaminated equipment.

An example of anthropometrics and human strength problems is the lack of easy access to equipment requiring maintenance. Two examples of the problems belonging to environmental factors are heat stress and a high variability of illumination. The communications problems include inadequate capacity of the existing communications system for satisfying the volume of communications traffic needed throughout the plant, particularly during outages, the protective clothing worn by maintenance personnel while working in radioactive environment causes serious impediments to effective communications and insufficient communication coverage throughout the power plant. Finally, some examples of personnel safety problems are steam burns, heat prostration, radiation exposure, and chemical burns.

Additional information on all of the above groups of findings is available in References 6 and 7.

10.3 Elements Relating to Human Performance That Can Contribute to an Effective Maintenance Program in Nuclear Power Plants

There are many elements relating to human performance that can directly or indirectly contribute to an effective maintenance program in nuclear power plants. Six of these elements are as follows [3,8]:

- **Element 1:** Good planning
- **Element 2:** Design for maintainability

- **Element 3:** Provision of usable, accurate, and up-to-date maintenance procedures
- **Element 4:** Human performance program, including error prevention tools, and performance measures
- **Element 5:** Strategies to learn from past experiences
- **Element 6:** Provision of appropriately trained and experienced maintenance personnel

Additional information on the above elements is available in Reference 8.

10.4 Useful Human Factors Methods to Assess and Improve Nuclear Power Plant Maintainability

There are many human factors methods that can be used to assess and improve nuclear power plant maintainability. Five of these methods are described below, separately [6,9].

10.4.1 Structured Interviews

This is one of the most effective methods used for collecting valuable data concerning maintainability in the shortest possible time. The method assumes that personnel such as repair persons, technicians, and their supervisors close to maintainability-related problems generally provide the most meaningful insights into the difficulties involved in carrying out their job the best possible way. In a structured interview, a fixed set of questions such as presented below are asked [6,9,10].

- Are appropriate lay down areas and workbenches provided?
- Is your workshop facility arranged in such a way so that it allows efficient and safe performance of all maintenance-related activities?
- How would you describe the environment in your workshop facility with regard to factors such as noise, illumination, and ventilation?
- Is our workshop facility sized properly for accommodating effectively all the personnel in your organization?
- How well is your workshop facility integrated into the overall/total plant design?

After analyzing all the collected data, necessary recommendations for improvements are made. Additional information on this method is available in References 6 and 9.

10.4.2 Task Analysis

This is a systematic approach used for assessing the equipment maintainer's requirements for successfully working with hardware to accomplish a specified task. The involved analyst oversees and records each task element and start and completion times in addition to making observations that are impediments to effective maintainability.

The observations are grouped under 16 classifications. These are equipment damage potential, availability of appropriate maintenance information (i.e., procedures, manuals, and schematics), environmental factors, equipment maintainability design features, facility design features, decision-making factors, access factors, workshop adequacy, personal hazards, lifting or movement aids, maintenance crew interactions, communication, tools and job aids, spare parts retrieval, supervisor-subordinate relationships, and training needs [6].

Additional information on this method is available in References 6 and 9.

10.4.3 Surveys

This method is used when the results obtained through the application of methods such as structured interviews and task analysis indicate a need for more detailed examination of certain maintainability-related factors. Two examples of such scenario are presented below [9,10]:

- **Example 1:** Most maintenance personnel have expressed concerns in the area of communications. Under such situations, it might be quite helpful to carry out a survey or test of message intelligibility between important communication links within the power plant facility.

- **Example 2:** Inadequate illumination is, directly or indirectly, proving to be a problem in the course of analyzing one or more certain tasks. In such situations, it might be quite helpful to carry out a power plant–wide illumination survey of all maintenance worksites.

Additional information on this method is available in Reference 9.

10.4.4 Potential Accident/Damage Analysis

This is a quite useful structured approach for assessing the accident, damage, or potential error inherent in a stated task. To determine the potential for mishaps' occurrence in the performance of a maintenance job, the starting point is to establish an appropriate mechanism that clearly describes the job under consideration in detail. Subsequently for each and every task element, the interviewer of the interviewee (e.g., repair person) asks the following question:

- Is there a low, medium, or high potential for the occurrence of an error/an accident/damage to system/equipment in carrying out, say, step xyz?

After analyzing all the collected data, appropriate changes to items such as equipment, facility, and procedures are recommended. Additional information on this method is available in References 6 and 9.

10.4.5 Critical Incident Technique

Past experiences over the years clearly indicate that maintenance errors', accidents', or near mishaps' history can provide useful information concerning needed maintainability-related improvements. In this regard, the critical incident technique is considered a very good tool for examining such case histories from the standpoint of human factors. The application of this tool/ technique calls for making appropriate arrangements to meet individually with members of the maintenance organization. To each individual, the following three questions are asked:

- Give one example of a plant system or unit of equipment that is quite well "human engineered" or quite straightforward to maintain, and describe the unit/system by emphasizing the features that make it good from the maintainer's perspective.
- Based on your personal experience, give one example of a maintenance error, accident, or near mishap with serious or potentially serious consequences. Furthermore, describe the specifics of the case and indicate the possible ways the situation could have been averted.
- Give one example of a plant system or unit of equipment that is not properly "human engineered" or from the maintenance person's perspective is quite poorly designed and which has resulted in or could lead to a safety hazard, damage to equipment, or an error.

After analyzing all the collected data, necessary changes for improvements are recommended. Additional information on this technique is available in References 6 and 9.

10.5 Nuclear Power Plant Maintenance Error-Related Facts, Figures, and Examples

Some of the facts, figures, and examples that are directly or indirectly concerned with human error in nuclear power plant maintenance are as follows:

- In 1990, a study of 126 human error-related significant events in the area of nuclear power generation reported that approximately 42% of the problems were linked to maintenance and modification [11].

- As per References 12 and 13, a study of over 4400 maintenance history records covering the period from 1992–1994 concerning a boiling water reactor (BWR) nuclear power plant revealed that approximately 7.5% of all failure records could be categorized as human errors related to maintenance activities.
- In 1989 on Christmas Day in the state of Florida, two nuclear reactors were shut down due to maintenance error and caused rolling blackouts [14].
- As per Reference 5, a study of nuclear power plant operating-related experiences revealed that due to errors in maintenance of some motors in the rod drives, many of these motors ran in a backward direction and withdrew rods instead of inserting them.
- As per Reference 15, in South Korean nuclear power plants, about 25% of sudden shutdowns were due to human errors, out of which over 80% were human errors resulting from normal testing and maintenance-related tasks.
- As per Reference 16, a study of 199 human errors that took place in Japanese nuclear power plants during the period from 1965–1995 reported that approximately 50 of them were concerned with maintenance activities.

10.6 Causes of Human Error in Nuclear Power Plant Maintenance and Maintenance Tasks Most Susceptible to Human Error in Nuclear Power Plants

There are many causes for the human error occurrence in nuclear power plant maintenance activities. On the basis of characteristics obtained from modelling the maintenance-related task, error causes in nuclear power plant maintenance may be grouped under the following four classifications [4]:

- **Design-related shortcomings in software and hardware.** These shortcomings include items such as confusing or wrong procedures, insufficient communication equipment, and deficiencies in the design of displays and controls.
- **Human ability-related limitations.** An example of these limitations is the limited capacity of short-term memory in the internal control mechanism.
- **Induced circumstances.** These circumstances include items such as emergency conditions; improper communications, which may lead to failures; and momentary distractions.

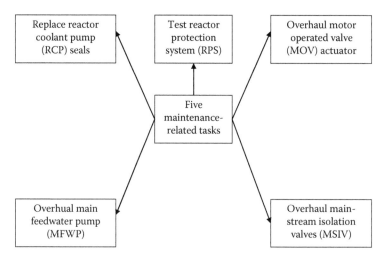

FIGURE 10.1
Five maintenance-related tasks most susceptible to human errors in nuclear power plants.

- **Disturbances of the external environment.** Some important examples of these disturbances are the physical conditions such as humidity, temperature, ambient illumination, and ventilation.

The Electric Power Research Institute (EPRI) in the United States and the Central Research Institute of Electric Power Industry (CRIEPI) in Japan carried out a joint study in the 1990s to identify critical maintenance-related tasks and to develop, implement, and evaluate interventions that have a very high potential for reducing the occurrence of human errors or increasing maintenance productivity in nuclear power stations. As the result of this joint study, five maintenance-related tasks most susceptible to the human error occurrence, as shown in Figure 10.1, were highlighted [17].

Additional information on these five maintenance-related tasks is available in Reference 17.

10.7 Digital Plant Protection Systems Maintenance Task-Related Human Errors

Nowadays, in order to take advantage of digital technology, the analog test reactor protection systems are being replaced by the digital plant protection systems. The scope of human error occurrence incidents in digital plant protection systems during the performance of a maintenance-related task is quite high, ranging from missing an important step of work procedure to intentional deviation of work procedure from the proper work procedure in order to save time or accomplishing the task easily in an uncomfortable environment.

The types of human errors that can occur in digital plant protection systems maintenance tasks are as follows [18,19]:

- **Resetting error.** This type of error occurs from a failure to reset bistable process parameters after completion of a test.
- **Calibration error.** This type of error is associated with an incorrect setting of trip limits or references.
- **Installation/repair error.** This type of error occurs when faulty parts are replaced or repaired during the refueling maintenance process as a corrective or preventive measure.
- **Quality error.** This type of error occurs basically due to carelessness and limited space for work or transport. Two typical examples of quality error are deficient welding/soldering joints or insulation and too little or too much tightening of screws.
- **Bypass error.** This type of error occurs whenever a channel is bypassed to conduct tests in that very channel.
- **Restoration error.** This type of error occurs from an oversight to restore the system after completion of maintenance or a test.

10.8 Useful Guidelines for Human Error Reduction and Prevention in Nuclear Power Plant Maintenance

Over the years, various useful guidelines have been proposed for reducing and preventing human error occurrence in nuclear power plant maintenance. Four of these guidelines are presented below [4,10]:

- **Guideline 1: Ameliorate design-related deficiencies.** This guideline calls for overcoming deficiencies in areas such as plant layout, labeling, work environment, and coding.
- **Guideline 2: Develop proper work safety checklists for maintenance personnel.** This guideline calls for providing maintenance personnel safety checklists that can be used to determine the possibility of human error occurrence as well as the factors that may affect their actions before or after the performance of maintenance tasks.
- **Guideline 3: Revise training programs for all involved maintenance personnel.** This guideline calls for training programs for all

involved maintenance personnel to be revised in accordance with the characteristics and frequency of occurrence of each extrinsic cause.

- **Guideline 4: Carry out administrative policies more thoroughly.** This guideline calls for motivating maintenance personnel appropriately to comply with prescribed quality control-related procedures.

Additional information on the above guidelines is available in Reference 4.

10.9 Methods for Performing Maintenance Error Analysis in Nuclear Power Plants

There are many methods/models that can be used for performing human error analysis in the area of nuclear power plant maintenance. Three of these methods/models are presented below.

10.9.1 Markov Method

This method is widely used for performing probability analysis of repairable engineering systems, and it can also be used for performing human error analysis in the area of nuclear power plant maintenance. The method is described in Chapter 4, and its application to perform maintenance error analysis in nuclear power plants is demonstrated through the mathematical model presented below.

This mathematical model represents a repairable nuclear power plant system that may fail due to a maintenance error or nonmaintenance error failures. The nuclear power plant system state space diagram is shown in Figure 10.2 [10,20]. Numerals in the circle and boxes denote system states. The following assumptions are associated with the mathematical model:

- The nuclear power plant system maintenance error and nonmaintenance error failure rates are constant.

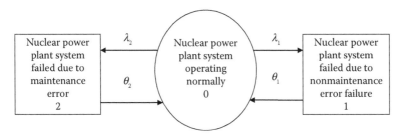

FIGURE 10.2
Nuclear power plant system state space diagram.

- The failed nuclear power plant system repair rates are constant.
- The repaired nuclear power plant system is as good as new.

The following symbols are associated with the mathematical model:

j is the nuclear power plant system state j; for $j = 0$ (nuclear power plant system operating normally), $j = 1$ (nuclear power plant system failed due to nonmaintenance error failure), $j = 2$ (nuclear power plant system failed due to maintenance error).

$P_j(t)$ is the probability that the nuclear power plant system is in state j at time t; for $j = 0, 1, 2$.

λ_1 is the nuclear power plant system constant nonmaintenance error failure rate.

λ_2 is the nuclear power plant system constant maintenance error rate.

θ_1 is the nuclear power plant system constant repair rate from sate 1 to state 0.

θ_2 is the nuclear power plant system constant repair rate from state 2 to state 0.

By using the Markov method described in Chapter 4, we write the following equations for Figure 10.2 state space diagram:

$$\frac{dP_0(t)}{dt} + (\lambda_1 + \lambda_2)P_0(t) = \theta_1 P_1(t) + \theta_2 P_2(t) \tag{10.1}$$

$$\frac{dP_1(t)}{dt} + \theta_1 P_1(t) = \lambda_1 P_0(t) \tag{10.2}$$

$$\frac{dP_2(t)}{dt} + \theta_2 P_2(t) = \lambda_2 P_0(t) \tag{10.3}$$

At time $t = 0$, $P_0(0) = 1$, $P_1(0) = 0$, and $P_2(0) = 0$.
By solving Equations 10.1 through 10.3, we get

$$P_0(t) = \frac{\theta_1\theta_2}{y_1 y_2} + \left[\frac{(y_1 + \theta_2)(y_2 + \theta_1)}{y_1(y_1 - y_2)}\right]e^{y_1 t} - \left[\frac{(y_2 + \theta_2)(y_2 + \theta_1)}{y_2(y_1 - y_2)}\right]e^{y_2 t} \tag{10.4}$$

where

$$y_1, y_2 = \frac{B \pm [B^2 - 4(\theta_1\theta_2 + \lambda_2\theta_1 + \lambda_1\theta_1)]^{1/2}}{2} \tag{10.5}$$

$$B = \theta_1 + \theta_2 + \lambda_1 + \lambda_2 \tag{10.6}$$

$$y_1 y_2 = \theta_1 \theta_2 + \lambda_2 \theta_1 + \lambda_1 \theta_2 \tag{10.7}$$

$$y_1 + y_2 = -(\theta_1 + \theta_2 + \lambda_1 + \lambda_2) \tag{10.8}$$

$$P_1(t) = \frac{\lambda_2 \theta_1}{y_1 y_2} + \left[\frac{\lambda_1 y_1 + \lambda_1 \theta_2}{y_1(y_1 - y_2)} \right] e^{y_1 t} - \left[\frac{(\theta_2 + y_2)\lambda_2}{y_2(y_1 - y_2)} \right] e^{y_2 t} \tag{10.9}$$

$$P_2(t) = \frac{\lambda_1 \theta_2}{y_1 y_2} + \left[\frac{\lambda_1 y_1 + \lambda_1 \theta_2}{y_1(y_1 - y_2)} \right] e^{y_1 t} - \left[\frac{(\theta_2 + y_2)\lambda_1}{y_2(y_1 - y_2)} \right] e^{y_2 t} \tag{10.10}$$

As t becomes very large, we get the following steady-state probability equations from Equations 10.4, 10.9, and 10.10, respectively:

$$P_0 = \frac{\theta_1 \theta_2}{y_1 y_2} \tag{10.11}$$

$$P_1 = \frac{\lambda_2 \theta_1}{y_1 y_2} \tag{10.12}$$

$$P_2 = \frac{\lambda_1 \theta_2}{y_1 y_2} \tag{10.13}$$

where

P_0, P_1, P_2 are the steady-state probabilities of the nuclear power plant system being in states 0, 1, and 2, respectively.

It should be noted that Equation 10.11 is also known as the system steady-state availability. In this case, it is the steady-state availability of the nuclear power plant system.

EXAMPLE 10.1

Assume that for a nuclear power plant system, we have the following data values

$\lambda_1 = 0.0008$ failures per hour
$\theta_1 = 0.06$ repairs per hour
$\lambda_2 = 0.0001$ errors per hour
$\theta_2 = 0.04$ repairs per hour

Calculate the nuclear power plant system steady-state availability and the steady-state probability of failing due to maintenance error.

By substituting the given data values into Equation 10.11, we get

$$P_0 = \frac{\theta_1\theta_2}{y_1y_2} = \frac{\theta_1\theta_2}{\theta_1\theta_2 + \lambda_2\theta_1 + \lambda_1\theta_2}$$

$$= \frac{(0.06)(0.04)}{(0.06)(0.04) + (0.0001)(0.06) + (0.0008)(0.04)}$$

$$= 0.9844$$

Similarly, by inserting the specified data values into Equation 10.13, we get

$$P_2 = \frac{\lambda_1\theta_2}{y_1y_2} = \frac{\lambda_1\theta_2}{\theta_1\theta_2 + \lambda_2\theta_1 + \lambda_1\theta_2}$$

$$= \frac{(0.0008)(0.04)}{(0.06)(0.04) + (0.0001)(0.06) + (0.0008)(0.04)}$$

$$= 0.0131$$

Thus, the nuclear power plant system steady-state availability and the steady-state probability of failing due to maintenance error are 0.9844 and 0.0131, respectively.

10.9.2 Fault Tree Analysis

This method is often used to perform various types of reliability-related analysis in the industrial sector [21,22]. The method is described in detail in Chapter 4. Its application to perform human error analysis in the area of nuclear power plant maintenance is demonstrated through the example presented below.

EXAMPLE 10.2

Assume that a system used in a nuclear power plant can fail due to a maintenance error caused by any of these four factors: poor system design, carelessness, poor work environment, and use of deficient maintenance manuals. Two major factors for carelessness are time constraints or poor training. Similarly, two factors for a poor work environment are distractions or poor lighting.

Develop a fault tree for the top event "nuclear power plant system failure due to a maintenance error" by using fault tree symbols given in Chapter 4.

A fault tree for the example is shown in Figure 10.3.

EXAMPLE 10.3

Assume that the occurrence probability of events Y_1, Y_2, Y_3, Y_4, Y_5 and Y_6 shown in Figure 10.3 is 0.04. For independent events, calculate the

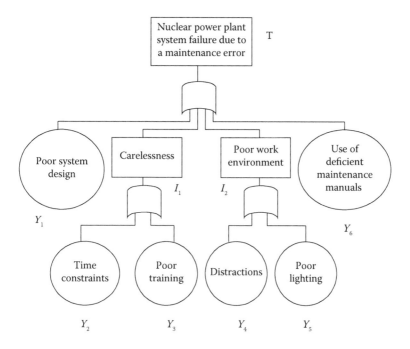

FIGURE 10.3
Fault tree for Example 10.2.

occurrence probability of the top event T (i.e., nuclear power plant system failure due to a maintenance error), and intermediate events I_1 (i.e., carelessness), and I_2 (i.e., poor work environment). Also, redraw Figure 10.3 fault tree with given and calculated values.

By using Chapter 4 and the given data values, we obtain the values of I_1, I_2 and T as follows:

The probability of occurrence of event I_1 is given by

$$P(I_1) = P(Y_2) + P(Y_3) - P(Y_2)P(Y_3)$$
$$= 0.04 + 0.04 - (0.04)(0.04)$$
$$= 0.0784$$

where

$P(I_1)$, $P(Y_2)$, and $P(Y_3)$ are the occurrence probabilities of events I_1, Y_2, and Y_3, respectively.

The probability of occurrence of event I_2 is given by

$$P(I_2) = P(Y_4) + P(Y_5) - P(Y_4)P(Y_5)$$
$$= 0.04 + 0.04 - (0.04)(0.04)$$
$$= 0.0784$$

where
    $P(I_2)$, $P(Y_4)$, and $P(Y_5)$ are the occurrence probabilities of events I_2, Y_4 and Y_5, respectively.

By using the above calculated and given data values and Chapter 4, we get

$$P(T) = 1 - \{1 - P(Y_1)\}\{1 - P(I_1)\}\{1 - P(I_2)\}\{1 - P(Y_6)\}$$
$$= 1 - (1 - 0.04)(1 - 0784)(1 - 0.0784)(1 - 0.04)$$
$$= 0.2172$$

where
    $P(T)$, $P(Y_1)$ and $P(Y_6)$ are the occurrence probabilities of events T, Y_1, and Y_6, respectively.

Thus, the occurrence probabilities of events T, I_1, and I_2 are 0.2172, 0.0784, and 0.0784, respectively. Figure 10.3 fault tree with specified and calculated fault event occurrence probability values is shown in Figure 10.4.

10.9.3 Maintenance Personnel Performance Simulation (MAPPS) Model

This is a computerized, stochastic, task-oriented human behavioral model. It was developed by the Oak Ridge National Laboratory to provide estimates of nuclear power plant maintenance manpower performance

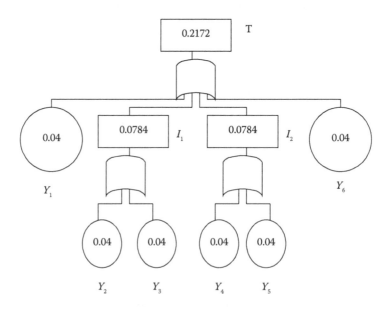

FIGURE 10.4
A fault tree with given and calculated fault occurrence probability values.

measures [23]. The development of this model was sponsored by the U.S. NRC, and the main objective for its development was the pressing need for and lack of a human reliability-related data bank pertaining to nuclear power plant maintenance activities for use in conducting probabilistic risk assessment studies [23].

Some of the performance measures estimated by the MAPPS model are identification of the most and least likely error-prone sub elements, probability of successfully accomplishing the task of interest, probability of an undetected error, the task duration time, and maintenance team stress profiles during task execution. Needless to say, the MAPPS model/method is an excellent tool to estimate important maintenance parameters, and its flexibility allows it to be useful for various applications concerned with nuclear power plant maintenance activity.

Additional information on this model/method is available in Reference 23.

10.10 Problems

1. Write an essay on human factors and human error in nuclear power plant maintenance.
2. Discuss at least four types of human factors engineering maintenance-related shortcomings/problems in nuclear power plant systems.
3. What are the elements relating to human performance that can contribute to an effective maintenance program in nuclear power plants?
4. Discuss at least three human factors methods for assessing and improving nuclear power plant maintainability.
5. Discuss at least five facts, figures, and examples that are directly or indirectly concerned with human error in nuclear power plant maintenance.
6. What are the main causes for the occurrence of human error in nuclear power plant maintenance?
7. What are the maintenance tasks most susceptible to human error in nuclear power plants?
8. Discuss useful guidelines for human error reduction and prevention in nuclear power plant maintenance.
9. Prove Equations 10.4, 10.9, and 10.10 by using Equations 10.1 through 10.3.
10. What are the types of human errors that can occur in digital plant protection system maintenance tasks?

References

1. Reactor Safety Study: An Assessment of Accident Risks in U.S. Commercial Nuclear Power Plants, Report No. WASH-1400, United States Nuclear Regulatory Commission, Washington, DC, 1975.
2. Seminara, J. L., Gonzalez, W.R., Parsons, S.O., Human Factors Review of Nuclear Power Plant Control Room Design, Report No. EPRI NP-309, Electric Power Research Institute (EPRI), Palo Alto, California, 1976.
3. Dhillon, B.S., *Human Reliability, Error, and Human Factors in Power Generation*, Springer, London, 2014.
4. Wu, T.M., Hwang, S.L., Maintenance error reduction strategies in nuclear power plants, using root cause analysis, *Applied Ergonomics*, Vol. 20, No. 2, 1989, pp. 115–121.
5. Nuclear Power Plant Operating Experiences, from the IAEA/NEA Incident Reporting System 1996–1999, Organization for Economic Co-operation and Development (OECD), Paris, 2000.
6. Seminara, J.L., Parsons, S.O., Human factors engineering and power plant maintenance, *Maintenance Management International*, Vol. 6, 1985, pp. 33–71.
7. Seminara, J.L., Parsons, S.O., Human Factors Review of Power Plant Maintainability, Report No. EPRI NP-5714, Electric Power Research Institute (EPRI), Palo Alto, California, 1981.
8. Penington, J., Shakeri, S., A human factors approach to effective maintenance, *Proceedings of the 27th Annual Canadian Nuclear Society Conference*, pp. 1–11.
9. Seminara, J.L., Human Factors Methods for Assessing and Enhancing Power Plant Maintainability, Report No. EPRI NP-2360, Electric Power Research Institute, Palo Alto, California, 1982.
10. Dhillon, B.S., *Human Reliability, Error, and Human Factors in Engineering Maintenance: with reference to aviation and power generation*, CRC Press, Boca Raton, Florida, 2009.
11. Reason, J., Human Factors in Nuclear Power Generation: A Systems Perspective, *Nuclear Europe Worldscan*, Vol. 17, No. 5–6, 1997, pp. 35–36.
12. Pyy, P., Laakso, K., Reimann, L., A study of human errors related to NPP maintenance activities, Proceedings of the IEEE 6th Annual Human Factors Meeting, 1997, pp. 12.23–12.28.
13. Pyy, P., An analysis of maintenance failures at a nuclear power plant, *Reliability Engineering and System Safety*, Vol. 72, 2001, pp. 293–302.
14. Maintenance Error a Factor in Blackouts, *Miami Herald*, Miami, Florida, December 29, 1989, pp. 4.
15. Heo, G., Park, J., Framework of quantifying human error effects in testing and maintenance, *Proceedings of the Sixth American Nuclear Society International Topical Meeting on Nuclear Plant Instrumentation, Control, and Human-Machine Interface Technologies*, 2009, pp. 2083–2092.
16. Hasegawa, T., Kameda, A., Analysis and Evaluation of Human Error Events in Nuclear Power Plants, Presented at the Meeting of the IAEA'S CRP on "Collection and Classification of Human Reliability Data for Use in Probabilistic Safety Assessments", May 1998. *Available from the Institute of Human Factors, Nuclear Power Engineering Corporation*, 13-17-1, Toranomon, Minato-ku, Tokyo, Japan.

17. Isoda, H., Yasutake, J.Y., Human factors interventions to reduce human errors and improve productivity in maintenance tasks, *Proceedings of the International Conference on Design and Safety of Advanced Nuclear Power Plants*, 1992, pp. 34.4-1–34.4-6.

18. Khalaquzzaman, M., Kang, H.G., Kim, M.C., Seong, P.H., A model for estimation of reactor spurious shutdown rate considering maintenance human errors inrReactor protection system of nuclear power plants, *Nuclear Engineering Design*, Vol. 240, 2010, pp. 2963–2971.

19. DPPS Maintenance Manual, Ulchin Nuclear Power Plant Unit No. 586, Document No. 9-650-Z-431-100, Westinghouse Corporation, Butler County, Pennsylvania, 2002.

20. Dhillon, B.S., *Engineering Maintenance: A Modern Approach*, CRC Press, Boca Raton, Florida, 2002.

21. Dhillon, B.S., Singh, C., *Engineering Reliability: New Techniques and Applications*, John Wiley & Sons, New York, 1981.

22. Dhillon, B.S., *Human Reliability: with Human Factors*, Pergamon Press, Inc., New York, 1986.

23. Knee, H.E., The maintenance personnel performance simulation (MAPPS) model: A human reliability analysis tool, *Proceedings of the International Conference on Nuclear Power Plant Aging, Availability Factor and Reliability Analysis*, 1985, pp. 77–80.

11

Human Factors in Nuclear Power Plant Control Systems

11.1 Introduction

In nuclear power plants, control systems play a very important role. Over the years, control systems and their operators' role have changed quite dramatically. The human operator's activity has evolved from manually carrying out the process to control system supervision. In turn, the human operator needs effective man-machine interfaces, an in-depth knowledge of the process being monitored, the ability for making effective decisions within demanding constraints, etc.

Needless to say, the human factors in control systems in the area of nuclear power generation have become a very important issue after the occurrence of the Three Mile Island Nuclear Power Plant accident in 1979. A subsequent study conducted by the U.S. Nuclear Regulatory Commission (NRC) reported the findings such as violation of number of human engineering-related principles in control panel designs; the information needed by operators was often nonexistent, ambiguous, difficult to read, or poorly located; and virtually nonexistent of human engineering at the Three Mile Island Nuclear Power Plant [1,2].

This chapter presents various important aspects of human factors in nuclear power plant control systems.

11.2 Human Performance-Related Advanced Control Room Technology Issues and Control Room Design-Related Deficiencies That Can Lead to Human Error

Past experiences over the years clearly indicate that although advanced technology has the potential for improving system performance, there is also potential for negatively impacting human performance and creating precursors for the occurrence of human errors. Ten of the human performance-related advanced control room technology issues are presented below [3–5]:

- **Issue 1:** Considerable loss of skill-related proficiency for the occasional performance of tasks that are generally automated.
- **Issue 2:** Shift from physical to rather highly cognitive inclined workload, impairing the operator's ability to properly monitor and process all types of relevant data.
- **Issue 3:** Considerably high shifts in operator workload (i.e., workload transition) whenever a computer failure takes place.
- **Issue 4:** Increment in operator's cognitive workload concerning the management of the interface (i.e., positioning, scaling, and opening windows).
- **Issue 5:** Operator's loss of vigilance due to automated systems resulting in decrease in the ability to detect off-normal conditions.
- **Issue 6:** Difficulties in navigating through and finding important information presented in a computer-based workspace.
- **Issue 7:** Ill-identified and poorly organized tasks resulting from function-related strategies' poor allocation.
- **Issue 8:** Loss of ability for utilizing well-learned, rapid eye-scanning patterns as well as recognition of patterns from spatially fixed parameter-related displays, particularly in the case of highly flexible CRT-type interfaces.
- **Issue 9:** Loss of operator's "situation awareness," which makes it quite difficult to assume direct control when necessary.
- **Issue 10:** Difficulty in comprehending how advanced systems function, which can result in either a lack of operator aids' complete acceptance or too much reliance on them.

Additional information on these 10 issues is available in Reference 3.

There are many nuclear poor plant control room design-related deficiencies that can result in human error. Fourteen of these deficiencies are as follows [6]:

- **Deficiency 1:** Chart recorders contain too many parameters.
- **Deficiency 2:** Poor labeling practice, including inconsistency in abbreviations.
- **Deficiency 3:** Controls and displays placed well beyond anthropometric reach and vision envelops.
- **Deficiency 4:** Poor location of some controls that can directly or indirectly result in advertent activation.

- **Deficiency 5:** Adjacent meters with different scales that must be compared by operators.
- **Deficiency 6:** Poor application of shape coding and mirror-imaged control boards.
- **Deficiency 7:** Glare and reflection from lighting on all involved instruments.
- **Deficiency 8:** Subsystem controls widely separated from their associated alarm annunciators.
- **Deficiency 9:** Highly complex annunciator systems along with complex equipment/procedures for knowledge, testing, resetting, and silencing alarms.
- **Deficiency 10:** Lack of appropriate barriers for control knobs or switches considered critical.
- **Deficiency 11:** Meters that fail with the pointer reading in the scale's normal band.
- **Deficiency 12:** Illegible printouts of recorders and use of qualitative instead of quantitative indicators.
- **Deficiency 13:** Inconsistency in color coding within a control room.
- **Deficiency 14:** Incorrect application of minor, intermediate, and major scale-related markings on involved meters.

11.3 Human Engineering Discrepancies in Control Room Visual Displays

A study of a control room survey of several nuclear power plants reported many human engineering-related discrepancies in control room visual displays [5,7]. The following were the main areas of these discrepancies [5,7]:

- **Scale markings:** Three discrepancies of this area were incorrect scale graduation marks, poor numerical progression, and log multiple-scales.
- **Color coding:** Two discrepancies of this area were inconsistent color coding practices and poor color usage.
- **Information displayed:** Two discrepancies of this area were incorrect display type and failed displays not apparent.
- **Scale zone markings:** Three discrepancies of this area were no scale banding, no alarm points, and informal banding.

- **Display readability:** Three discrepancies of this area were characteristics and marking too small, informal meter scales, and poor contrast (glare).
- **Usability of displays:** Two discrepancies of this area were inappropriate scale of scale range limits and conversion required.

11.4 Human Factor-Related Evaluation of Control Room Annunciators

In a nuclear power plant control room, a large number of annunciators are used for keeping control room operators alert. For example, a typical commercial nuclear power plant control room can have from 1000 to 2000 annunciators [8]. These annunciators are very important to system safety design as they are activated whenever plant parameter limits exceed. Thus, a study was carried out for identifying specific problem areas conducive to operator difficulty as well as for providing appropriate specific and generic solutions to these problems [8].

The study evaluated the annunciator systems used in four different nuclear reactor facilities and formally interviewed 39 reactor operators. A series of structured questions such as presented below were asked to each operator during the interview [5,8]:

- How useful are the annunciators for diagnosing plant conditions?
- Which annunciators in your opinion are most helpful/useful?
- How easy to use are the annunciators' procedures?
- Do you have any suggestions for improving annunciator systems, training, and the methods utilized by operators?
- Are any of the annunciator displays cumbersome to read or comprehend?
- What type of annunciator-associated hardware do you consider the most suitable?
- Are you fully satisfied with the existing locations of all the annunciators?
- How easy or difficult is it for maintaining the existing annunciators?
- Is there a situation for which no annunciator available?
- Have you ever had any past problems or certain areas of concern with any annunciator displays?

The final results of the information collected and analyzed were grouped under five classifications presented below, separately [8].

Classification 1: Operator Problems

- For effective viewing, labeling and letter sizes are quite unsatisfactory.
- Annunciators are often employed to display various types of unimportant or relatively minor conditions.
- In the situation when a major system failure takes place, many alarms that are only indirectly concerned with the primary failure are turned on. In turn, this quite significantly increases the information load that concerned operators must cognitively process for correctly highlighting the primary failure source.

Classification 2: Maintenance Problems
The items belonging to this classification are as follows:

- In some power plants, tag-out procedures for annunciator maintenance are very poor.
- Usually bulb and logic card-related repair is relatively quite often.
- In some alarm systems, operators are required to make their own tools to change bulbs.

Classification 3: Equipment Shortcomings
Six items belonging to this classification are as follows:

- Some annunciators are not properly equipped with an alarm override switch as set point adjustment that can be employed during maintenance for preventing constant and recurring acoustic alarms.
- Usually, there is no systematic correspondence between annunciator display elements and the controls employed for rectifying the alarms.
- Some control room annunciators have fairly low credibility with operators due to high number of false or misleading alarms.
- Usually, very little descriptive-related information is provided on window faces for orienting potential operators towards appropriate corrective actions.
- Some annunciators are not equipped with a press-to-test device for testing circuit and bulb.
- Annunciators do not fully employ existing computer display and logic capability for reducing or filtering the number of meaningless alarms, providing automatic default action if the operator fails to take appropriate action, helping the operator to focus on the fundamental cause of a given alarm, and providing coded or written procedures for operators so that proper corrective measures can be undertaken.

Classification 4: Lack of Organization and Consistency
The items belonging to this classification are as follows:

- A lack of uniformity in labeling, script, color coding, acoustic alarm frequency, abbreviations, timbre, contrast, design, and operating logic.
- Annunciators are quite rarely designed and placed in a consistent, rational, or logical manner.
- Annunciators are quite normally poorly organized and structured with regard to importance, function, system impact, or response immediacy.

Classification 5: Training Problems
Two items belonging to this classification are as follows:

- There is very little agreement among operators on what annunciators are most important or necessary for plant operations.
- Some guide-associated documentation is incomplete/inaccurate.

11.5 Benefits of Considering Human Factors in Digital Control Room Upgrades and an Approach for Incorporating Human Factor Considerations in Digital Control Room Upgrades

Past experiences over the years clearly indicate that there are many benefits of considering human factors in digital control room upgrades. These benefits may be grouped under four classifications shown in Figure 11.1 [9].

In the case of classification 1: *Increased operator acceptance,* potential operators can provide very important input to human factors design and evaluation. In the case of classification 2: *Increased plant availability,* operator controllable trips and inefficiencies can be avoided, and downtime for system testing and installation is shorter. Similarly, in the case of classification 3: *Improved equipment specification and procurement,* all necessary human factor-related considerations can be specified right from the start of the project.

Finally, in the case of classification 4: *Decreased need for back fits or redesigns,* all necessary changes can be carried out at earlier and less costly stages of the design process.

The approach for incorporating human factors considerations in digital control room upgrades is composed of the following five steps [5,9]:

- **Step 1:** This step is concerned with surveying the existing control room as well as its environment for determining requirements for the new computer-based workstations.

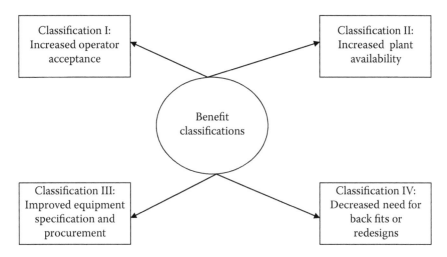

FIGURE 11.1
Classifications of benefits of considering human factors in digital control room upgrades.

- **Step 2:** This step is concerned with performing analysis of the tasks to be performed for developing information and control-related requirements.
- **Step 3:** This step is concerned with interviewing operators for obtaining input and feedback for gathering data from the end users early in the design process.
- **Step 4:** This step is concerned with reviewing preliminary drawings and mock-ups against human factor-related criteria at early stages of design to avoid costly back fits.
- **Step 5:** This step is concerned with developing display organization and control grouping strategies for applying to scenarios generated during the analysis for optimizing design trade-offs.

11.6 Recommendations for Overcoming Problems When Digital Control Room Upgrades Are Undertaken without Considering Human Factors into the Design

Recommendations for overcoming problems when digital control room upgrades undertaken without considering human factors in the design are grouped under six classifications as shown in Figure 11.2 [9]. These are displays, control devices, system hardware, facility, system software, and guidelines/training.

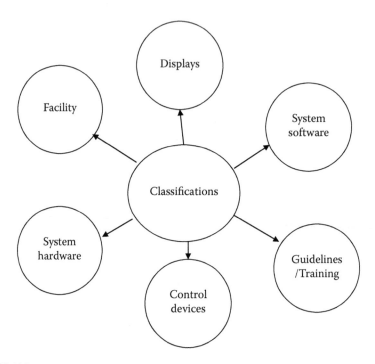

FIGURE 11.2
Classifications of recommendations for overcoming problems when digital control room upgrades undertaken without considering human factors into design.

The recommendations belonging to the displays classification are as follows [5,9]:

- Ensure display configuration consistency.
- Use color properly/effectively.
- Simplify all confusing displays.
- Include appropriate graphics.
- Improve summary-related displays to provide better/effective overall plant picture.
- Include all types of necessary data on relevant displays.
- Add displays considered relevant.

The recommendations belonging to the control devices classification are as follows [5,9]:

- Position all controls properly/effectively.
- Simplify all confusing control operations.

The recommendations belonging to the system hardware classification are as follows [5,9]:

- Ensure that equipment/system operation is not intrusive or distracting.
- Ensure that system hardware is appropriately designed to withstand operational-related rigors.
- Ensure that system hardware contains all types of necessary hardware controls.

The recommendations belonging to the facility classification are as follows [5,9]:

- Ensure that surrounding environment fully supports system operation and does not degrade it.
- Ensure that all system parts are properly placed in the surrounding environment.
- Ensure that system is properly configured so that the involved operator can effectively and easily carry out all required operations.

The recommendations belonging to the system software classification are as follows [5,9]:

- Minimize multiple and confusing steps required for obtaining data.
- Ensure that system performs all required operations appropriately and effectively.
- Optimize all involved update rates and response time/capacity.
- Ensure that system software effectively collects all types of relevant data.

Finally, the recommendations belonging to the guidelines/training classification are as follows [5,9]:

- Provide appropriate and effective guidelines for system implementation.
- Ensure appropriate and effective level of operator confidence through training.
- Provide appropriate and effective guidelines for display and control design and development.
- Provide appropriate and effective hands-on training.

11.7 Problems

1. Write an essay on human factors in nuclear power plant control systems.
2. Discuss at least eight human performance-related advanced control room technology issues.
3. List at least 14 control room design-related deficiencies that can lead to human error.
4. Discuss the main areas of human engineering discrepancies in control room visual displays.
5. Write an essay on human factor-related evaluation of control room annunciators.
6. What are the benefits of considering human factors in digital control room upgrades?
7. Describe an approach to incorporate human factors considerations in digital control room upgrades.
8. Discuss recommendations (in displays and control devices areas) for overcoming problems when digital control room upgrades are undertaken without considering human factors into design.
9. Discuss recommendations (in system hardware and facility areas) for overcoming problems when digital control room upgrades are undertaken without considering human factors in the design.
10. Discuss recommendations (in system software and guidelines/training areas) for overcoming problems when digital control room upgrades are undertaken without considering human factors into design.

References

1. Tennant, D.V., Human factors considerations in power plant control room design, *Proceedings of the 29th Power Instrumentation Symposium*, 1986, pp. 29–36.
2. Malone, T.B. et al., Human Factors Evaluation of Control Room Design and Operator Performance at Three Mile Island-2, Final Report, Report No. NUREG/CR-1270, Vol. 1, Nuclear Regulatory Commission, Washington, DC, 1980.
3. O'Hara, J.M., Hall, R.E., Advanced control rooms and crew performance issues: Implications for human reliability, *IEEE Transactions on Nuclear Science*, Vol. 39, No. 4, 1992, pp. 919–923.
4. Rasmussen, J., Duncan, K., Leplat, J., eds, *New Technology and Human Error*, John Wiley & Sons, New York, 1987.

5. Dhillon, B.S., *Human Reliability, Error, and Human Factors in Power Generation*, Springer International Publishing, Switzerland, 2014.
6. Brooks, D.M., Wilmington, C.R., Wilmington, T.G. Role of Human Factors in System Safety, *Proceedings of the International Congress on Advances in Nuclear Power Plants*, 2008, pp. 62–75.
7. Lisboa, J., Human factors assessment of digital monitoring systems for nuclear power plants control room, *IEEE Transactions on Nuclear Science*, Vol. 39, No. 4, 1992, pp. 9224–9232.
8. Banks, W.W., Blackman, H.S., Curtis, J.N., A human factors evaluation of nuclear power plant control room annunciators, *Proceedings of the Annual Control Engineering Conference*, 1984, pp. 323–326.
9. Gaddy, C.D., Fray, R.R., Divakaruni, S.M., Human factors considerations in digital control room upgrades, *Proceedings of the American Power Conference*, 1991, pp. 256–258.

12

Mathematical Models for Performing Safety, Reliability, and Human Error Analysis in Nuclear Power Plants

12.1 Introduction

In the area of engineering, mathematical modeling is a widely used approach for performing various types of analysis. In this case, the parts/components of a system are represented by idealized elements assumed to have representative characteristics of real-life parts/components, and whose behavior can be described by equations. However, the degree of realism of mathematical models very much depends on the assumptions imposed upon them.

Over the years, in the area of engineering safety and reliability engineering, a large number of mathematical models have been developed to study various aspects of safety, reliability, and human error in engineering systems. Most of these models were developed by using stochastic processes including the Markov method [1–4]. Although the effectiveness of such models can vary from one situation to another, some of these models are being used quite successfully to study various types of real-life environments in industry [5,6]. Thus, some of these models can also be used to study safety, reliability, and human error in nuclear power plants.

This chapter presents the mathematical models considered quite useful for performing various types of safety, reliability, and human error analysis in nuclear power plants.

12.2 Model I

This mathematical model represents a nuclear power plant system having three distinct states: working normally, failed safely, and failed unsafely. The safely and unsafely failed system is repaired back to its normal working

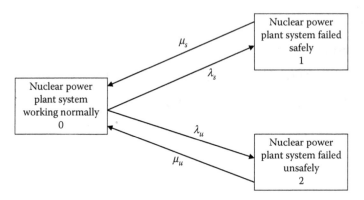

FIGURE 12.1
Nuclear power plant system state space diagram.

state. The nuclear power plant system state space diagram is shown in Figure 12.1 [7,8]. The numerals in the boxes denote the nuclear power plant system states.

The model is subjected to the following assumptions:

- The nuclear power plant system can fail either safely or unsafely.
- The nuclear power plant system safe and unsafe failure rates are constant.
- All failures occur independently.
- The failed nuclear power plant system repair rates are constant.
- The repaired nuclear power plant system is as good as new.

The following symbols are associated with the diagram/model:

i is the ith state of the nuclear power plant system, where $i = 0$ means the system is working normally, $i = 1$ means the system failed safely, and $i = 2$ means the system failed unsafely.

t is time.

$P_i(t)$ is the probability that the system is in state i at time t, for $i = 0, 1, 2$.

λ_i is the nuclear power plant system ith failure rate, where $i = s$ means safe and $i = u$ means unsafe.

μ_i is the failed nuclear power plant system's ith repair rate, where $i = s$ means from the safe failed state, and $i = u$ means from the unsafe failed state.

With the aid of the Markov method described in Chapter 4 and References 7 and 9, we write the following set of differential equations for Figure 12.1:

$$\frac{dP_0(t)}{dt} + (\lambda_s + \lambda_u)P_0(t) = \mu_s P_1(t) + \mu_u P_2(t) \tag{12.1}$$

$$\frac{dP_1(t)}{dt} + \mu_s P_1(t) = \lambda_s P_0(t) \tag{12.2}$$

$$\frac{dP_2(t)}{dt} + \mu_u P_2(t) = \lambda_u P_0(t) \tag{12.3}$$

At time $t = 0$, $P_0(0) = 1$ and $P_1(0) = P_2(0) = 0$.

By solving Equations 12.1 through 12.3, we obtain

$$P_0(t) = \frac{\mu_s \mu_u}{Z_1 Z_2} + \left[\frac{(Z_1 + \mu_s)(Z_1 + \mu_u)}{Z_1(Z_1 - Z_2)}\right]e^{Z_1 t} - \left[\frac{(Z_2 + \mu_s)(Z_2 + \mu_u)}{Z_2(Z_1 - Z_2)}\right]e^{Z_2 t} \tag{12.4}$$

where

$$Z_1, Z_2 = \frac{-A \pm \sqrt{A^2 - 4(\mu_s \mu_u + \lambda_s \mu_u + \lambda_u \mu_u)}}{2}$$

$$A = \mu_s + \mu_u + \lambda_u + \lambda_s$$

$$Z_1 Z_2 = \mu_u \mu_s + \lambda_s \mu_u + \lambda_u \mu_s$$

$$Z_1 + Z_2 = -(\mu_s + \mu_u + \lambda_s + \lambda_u)$$

$$P_1(t) = \frac{\lambda_s \mu_u}{Z_1 Z_2} + \left[\frac{(\lambda_s Z_1 + \lambda_s \mu_u)}{Z_1(Z_1 - Z_2)}\right]e^{Z_1 t} - \left[\frac{(\mu_u + Z_2)\lambda_s}{Z_2(Z_1 - Z_2)}\right]e^{Z_2 t} \tag{12.5}$$

$$P_2(t) = \frac{\lambda_u \mu_s}{Z_1 Z_2} + \left[\frac{(\lambda_u Z_1 + \lambda_u \mu_s)}{Z_1(Z_1 - Z_2)}\right]e^{Z_1 t} - \left[\frac{(\mu_s + Z_2)\lambda_u}{Z_2(Z_1 - Z_2)}\right]e^{Z_2 t} \tag{12.6}$$

Equations 12.5 and 12.6 give the probability of the nuclear power plant system failing safely and unsafely, respectively, when subjected to the repair process.

As time t becomes very large, the nuclear power plant system steady-state probability of failing safely using Equation 12.5 is

$$P_1 = \lim_{t \to \infty} P_1(t) = \frac{\lambda_s \mu_u}{Z_1 Z_2} \tag{12.7}$$

Similarly, as time t becomes very large, the nuclear power plant system steady-state probability of failing unsafely using Equation 12.6 is

$$P_2 = \lim_{t \to \infty} P_2(t) = \frac{\lambda_u \mu_s}{Z_1 Z_2} \tag{12.8}$$

By setting $\mu_s = \mu_u = 0$ in Equations 12.4 through 12.6 and then solving the resulting equations, we get

$$P_0(t) = e^{-(\lambda_s + \lambda_u)t} \tag{12.9}$$

$$P_1(t) = \frac{\lambda_s}{\lambda_s + \lambda_u} \left[1 - e^{-(\lambda_s + \lambda_u)t} \right] \tag{12.10}$$

$$P_2(t) = \frac{\lambda_u}{\lambda_s + \lambda_u} \left[1 - e^{-(\lambda_s + \lambda_u)t} \right] \tag{12.11}$$

Equation 12.9 gives the nuclear power plant system reliability at time t. In contrast, Equations 12.10 and 12.11 give the probability of the nuclear power plant system failing safely and unsafely at time t, respectively.

By integrating Equation 12.9 over the time interval $[0,\infty]$, we get the following equation for the nuclear power plant system mean time to failure [4,7]:

$$MTTF_{nps} = \int_0^\infty e^{-(\lambda_s + \lambda_u)t} dt$$

$$= \frac{1}{\lambda_s + \lambda_u} \tag{12.12}$$

where
$MTTF_{nps}$ is the nuclear power plan system mean time to failure.

EXAMPLE 12.1

Assume that a nuclear power plant system can fail safely or unsafely and its constant failure rates are 0.0008 failures per hour and 0.0002 failures per hour, respectively. Calculate the probability of the nuclear power plant system failing safely during a 200-hour mission and its mean time to failure.

By inserting the given data values into Equation 12.10, we obtain

$$P_1(200) = \frac{0.0008}{(0.0008 + 0.0002)} \left[1 - e^{-(0.0008 + 0.0002)(200)} \right]$$

$$= 0.1450$$

Similarly, by inserting the given data values into Equation 12.12, we get

$$MTTF_{nps} = \frac{1}{0.0008 + 0.0002}$$
$$= 1000 \text{ hours}$$

Thus, the probability of the nuclear power plant system failing safely during the specified mission period is 0.1450, and its mean time to failure is 1000 hours.

12.3 Model II

This mathematical model represents a nuclear power plant system having three states: working normally, working unsafely, and failed. The system is repaired from an unsafe working state and failed state, and its state space diagram is shown in Figure 12.2. The numerals in boxes denote nuclear power plant system states.

The following assumptions are associated with the diagram/model:

- The nuclear power plant system failure and repair rates are constant.
- The repaired nuclear power plant system is as good as new.
- All failures occur independently.

The following symbols are associated with this model:

i is the ith state of the nuclear power plant system, where $i = 0$ means the nuclear power plant system is working normally, $i = 1$ means the nuclear power plant system is working unsafely, and $i = 2$ the nuclear power plant system failed.

t is time.

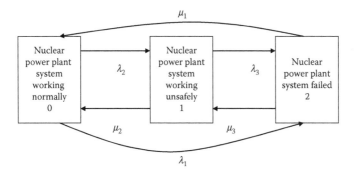

FIGURE 12.2
Nuclear power plant system state space diagram.

$P_i(t)$ is the probability that the nuclear power plant system is in state i at time t.

λ_i is the nuclear power plant system ith constant failure rate, where $i = 1$ means from state 0 to state 2, $i = 2$ means from state 0 to state 1, and $i = 3$ means from state 1 to state 2.

μ_i is the nuclear power plant system ith constant repair rate, where $i = 1$ means from state 2 to state 0, $i = 2$ means from state 1 to state 0, and $i = 3$ means from state 2 to state 1.

With the aid of the Markov method described in Chapter 4 and References 7 and 9, we write the following set of differential equations for Figure 12.2:

$$\frac{dP_0(t)}{dt} + (\lambda_1 + \lambda_2)P_0(t) = \mu_2 P_1(t) + \mu_1 P_2(t) \tag{12.13}$$

$$\frac{dP_1(t)}{dt} + (\mu_2 + \lambda_3)P_1(t) = \mu_3 P_2(t) + \lambda_2 P_0(t) \tag{12.14}$$

$$\frac{dP_2(t)}{dt} + (\mu_1 + \mu_3)P_2(t) = \lambda_3 P_1(t) + \lambda_1 P_0(t) \tag{12.15}$$

At time $t = 0$, $P_0(0) = 1$, and $P_1(0) = P_2(0) = 0$.

For a very large t, by solving Equations 12.13 through 12.15, we get the following steady-state probability equations [4,8]:

$$P_0 = \frac{(\mu_1 + \mu_3)(\mu_2 + \lambda_3) - \lambda_3 \mu_3}{D} \tag{12.16}$$

where

$$D = (\mu_1 + \mu_3)(\mu_2 + \lambda_2 + \lambda_3) + \lambda_1(\mu_2 + \lambda_3) + \lambda_1 \mu_3 + \lambda_2 \lambda_3 - \lambda_3 \mu_3$$

$$P_1 = \frac{\lambda_2(\mu_1 + \mu_3) + \lambda_1 \mu_3}{D} \tag{12.17}$$

$$P_2 = \frac{\lambda_1 \lambda_3 + \lambda_1(\mu_2 + \lambda_3)}{D} \tag{12.18}$$

where
P_0, P_1, and P_2 are the steady-state probabilities of the nuclear power plant system being in states 0, 1, and 2, respectively.

EXAMPLE 12.2

Assume that the following values of constant failure and repair rates of a nuclear power plant system are given:

$\lambda_1 = 0.006$ failures per hour
$\lambda_2 = 0.003$ failures per hour
$\lambda_3 = 0.001$ failures per hour
$\mu_1 = 0.008$ repairs per hour
$\mu_2 = 0.004$ repairs per hour
$\mu_3 = 0.009$ repairs per hour

Calculate the steady-state probability of the nuclear power plant system working unsafely with the aid of Equation 12.17.
By inserting the given data values into Equation 12.17, we obtain

$$P_1 = \frac{(0.008 + 0.009)(0.004 + 0.001) - (0.001)(0.009)}{D}$$

$$= 0.3551$$

where

$$D = (0.008 + 0.009)(0.004 + 0.003 + 0.001) + (0.006)(0.004 + 0.001)$$
$$+ (0.006)(0.009) + (0.003)(0.001) - (0.001)(0.009)$$

Thus, the steady-state probability of the nuclear power plant system working unsafely is 0.3551.

12.4 Model III

This mathematical model is concerned with predicting the reliability of a worker performing a task in a nuclear power plant under normal conditions—more specifically, the probability of performing a time-continuous task correctly by a worker. An equation for predicting the worker performance reliability is developed below [1,2,6,10].
The probability of human error in a task, in a nuclear power plant, in the finite time interval Δt with event B given is expressed by

$$P(A/B) = h(t)\Delta t \tag{12.19}$$

where
A is an event in which human error will occur in time interval $[t, t + \Delta t]$.
B is an errorless performance event of duration t.
$h(t)$ is the human error rate at time t.

The joint probability of the errorless performance is expressed by

$$P(\bar{A}/B) = P(B) - P(A/B)P(B) \tag{12.20}$$

where
$P(B)$ is the probability of occurrence of event B.
$\bar{A}$ is the event that human error will not occur in time interval $[t, t + \Delta t]$.

Equation 12.20 represents an errorless performance probability over time intervals $[0,t]$ and $[t, t + \Delta t]$ and is rewritten as

$$R_w(t) - R_w(t)P(A/B) = R_w(t + \Delta t) \tag{12.21}$$

where
$R_w(t)$ is the nuclear power plant worker reliability at time t.
$R_w(t + \Delta t)$ is the nuclear power plant worker reliability at time $(t + \Delta t)$.

By inserting Equation 12.19 into Equation 12.21, we obtain

$$\frac{R_w(t + \Delta t) - R_w(t)}{\Delta t} = -R_w(t)h(t) \tag{12.22}$$

In the limiting case Equation 12.22 becomes

$$\lim_{\Delta t \to 0} \frac{R_w(t + \Delta t) - R_w(t)}{\Delta t} = \frac{d R_w(t)}{dt} = -R_w(t)h(t) \tag{12.23}$$

At time $t = 0, R_w(0) = 1$.

By rearranging Equation 12.23, we obtain

$$\frac{1}{R_w(t)} \cdot dR_w(t) = -h(t)dt \tag{12.24}$$

By integrating both sides of Equation 12.24 over the time interval $[0,t]$, we obtain

$$\int_1^{R_w(t)} \frac{1}{R_w(t)} \cdot dR_w(t) = -\int_0^t h(t)dt \tag{12.25}$$

After evaluating Equation 12.25, we get

$$R_w(t) = e^{-\int_0^t h(t)dt} \tag{12.26}$$

Equation 12.26 is the general equation for calculating nuclear power plant worker performance reliability for any time to human error probability distribution (e.g., exponential, Weibull, and normal).

By integrating Equation 12.26 over the time interval $[0,\infty]$, we obtain the following general equation for the mean time to human error [4,11]:

$$MTTHE_{nw} = \int_0^\infty \left[e^{\int_0^\infty h(t)dt} \right] dt \tag{12.27}$$

where

$MTTHE_{nw}$ is the mean time to human error of a nuclear power plant worker.

EXAMPLE 12.3

Assume that a worker in a nuclear power plant is carrying out a certain task and his/her error rate is 0.008 errors/hour (i.e., his/her times to human error are exponentially distributed). Calculate the worker's reliability during an eight-hour work period.

Thus, we have

$h(t) = 0.008$ errors/hour

By inserting the above value and the specified value for time t into Equation 12.26, we obtain

$$R_w(6) = e^{-\int_0^8 (0.008)dt}$$
$$= e^{-(0.008)(8)}$$
$$= 0.9380$$

Thus, the nuclear power plant worker's reliability during the eight-hour work period is 0.9380.

12.5 Model IV

This mathematical model represents a nuclear power plant system that can fail due to human errors made by workers or due to hardware failures. The state space diagram of the model is shown in Figure 12.3. The numerals in the box and circles denote system states. It should be noted that mathematically this model is the special case of Model I, but its application is different.

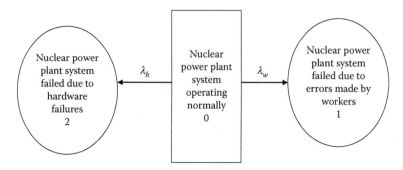

FIGURE 12.3
Nuclear power plant system state space diagram.

The following assumptions are associated with this model:

- Human errors and hardware failures occur independently.
- Both human error and hardware failure rates are constant.

The following symbols are associated with the diagram/model:

λ_w is the constant human error rate of the nuclear power plant workers.

λ_h is the constant hardware failure rate of the nuclear power plant system.

i is the ith state of the nuclear power plant system: $i = 0$ (nuclear power plant system operating normally), $i = 1$ (nuclear power plant system failed due to errors made by workers), $i = 2$ (nuclear power plant system failed due to hardware failures).

$P_i(t)$ is the probability of the nuclear power plant system being in state i at time t, for $i = 0, 1, 2$.

With the aid of the Markov method, we write the following equations for the Figure 12.3 diagram [2,4]:

$$\frac{dP_0(t)}{dt} + (\lambda_w + \lambda_h)P_0(t) = 0 \tag{12.28}$$

$$\frac{dP_1(t)}{dt} - \lambda_w P_0(t) = 0 \tag{12.29}$$

$$\frac{dP_2(t)}{dt} - \lambda_h P_0(t) = 0 \tag{12.30}$$

At $t = 0$, $P_0(0) = 1$, $P_1(0) = 0$, and $P_2(0) = 0$.

By solving Equations 12.28 through 12.30, we obtain

$$P_0(t) = e^{-(\lambda_w + \lambda_h)t} \tag{12.31}$$

$$P_1(t) = \frac{\lambda_w}{\lambda_w + \lambda_h}\left[1 - e^{-(\lambda_w + \lambda_h)t}\right] \tag{12.32}$$

$$P_2(t) = \frac{\lambda_h}{\lambda_w + \lambda_h}\left[1 - e^{-(\lambda_w + \lambda_h)t}\right] \tag{12.33}$$

The nuclear power plant system reliability is expressed by

$$R_{nps}(t) = P_0(t) = e^{-(\lambda_w + \lambda_h)t} \tag{12.34}$$

where

$R_{nps}(t)$ is the nuclear power plant reliability at time t.
The nuclear power plant system mean time to failure is given by

$$\begin{aligned} MTTF_{nps} &= \int_0^\infty R_{nps}(t)dt \\ &= \int_0^\infty e^{-(\lambda_w + \lambda_h)t}dt \\ &= \frac{1}{\lambda_w + \lambda_h} \end{aligned} \tag{12.35}$$

where

$MTTF_{nps}$ is the nuclear power plant system mean time to failure.

EXAMPLE 12.4

Assume that a nuclear power plant system can fail either due to hardware failures or human errors made by nuclear power plant workers. The system constant hardware failure and human error rates are 0.009 failures/hour and 0.002 errors/hour, respectively.

Calculate the probability that the system will fail due to a human error made by nuclear power plant workers during an eight-hour mission.

By inserting the given data values into Equation 12.32, we get

$$\begin{aligned} P_1(8) &= \frac{0.002}{(0.002 + 0.009)}\left[1 - e^{-(0.002 + 0.009)(8)}\right] \\ &= 0.0153 \end{aligned}$$

Thus, the probability that the system will fail due to a human error made by nuclear power plant workers is 0.0153.

12.6 Model V

This mathematical model represents a nuclear power plant worker performing time-continuous tasks in a fluctuating environment (i.e., normal and stressful). As the rate of human errors' occurrence can vary quite significantly from a normal work environment to a stressful work environment, this mathematical model can be used for calculating the nuclear power plant worker's reliability, probabilities of making error in stressful and normal environments, and mean time to human error in the fluctuating environment.

The model's state space diagram is shown in Figure 12.4, and the numerals in the diagram circles and boxes denote the state of the nuclear power plant worker.

The model is subjected to the following assumptions:

- Nuclear power plant worker errors occur independently.
- Nuclear power plant worker error rates in stressful and normal environments are constant.
- Rates of the environment changing from stressful to normal and vice versa are constant.

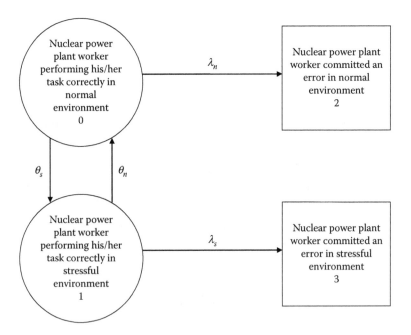

FIGURE 12.4
State space diagram for the nuclear power plant worker carrying out his/her task in fluctuating normal and stressful environments.

The following symbols are associated with this model:

j is the jth state of the nuclear power plant worker; $j = 0$ (nuclear power plant worker performing his/her task correctly in normal environment), $j = 1$ (nuclear power plant worker performing his/her task correctly in stressful environment), $j = 2$ (nuclear power plant worker committed an error in normal environment), $j = 3$ (nuclear power plant worker committed an error in stressful environment).

$P_j(t)$ is the probability of the nuclear power plant worker being in state j at time t, for $j = 0, 1, 2, 3$.

λ_n is the constant error rate of the nuclear power plant worker performing his/her task in a normal environment.

λ_s is the constant error rate of the nuclear power plant worker performing his/her task in a stressful environment.

θ_n is the constant transition rate from stressful work environment to normal work environment.

θ_s is the constant transition rate from normal work environment to stressful work environment.

By using the Markov method, we write the following set of differential equations for the Figure 12.4 diagram [4,11,12]:

$$\frac{dP_0(t)}{dt} + (\lambda_n + \theta_s)P_0(t) = \theta_n P_1(t) \tag{12.36}$$

$$\frac{dP_1(t)}{dt} + (\lambda_s + \theta_n)P_1(t) = \theta_s P_0(t) \tag{12.37}$$

$$\frac{dP_2(t)}{dt} = \lambda_n P_0(t) \tag{12.38}$$

$$\frac{dP_3(t)}{dt} = \lambda_s P_1(t) \tag{12.39}$$

At time $t = 0$, $P_0(0) = 1$, and $P_1(0) = P_2(0) = P_3(0) = 0$.

By solving Equations 12.36 through 12.39, we get the following state probability equations:

$$P_0(t) = \frac{1}{(x_1 - x_2)}\left[(x_2 + \lambda_s + \theta_n)e^{x_2 t} - (x_1 + \lambda_s + \theta_n)e^{x_1 t}\right] \tag{12.40}$$

where

$$x_1 = \left[-b_1 + \left(b_1^2 - 4b_2\right)^{1/2} \right] / 2 \tag{12.41}$$

$$x_2 = \left[-b_1 - \left(b_1^2 - 4b_2\right)^{1/2} \right] / 2 \tag{12.42}$$

$$b_1 = \lambda_n + \lambda_s + \theta_n + \theta_s \tag{12.43}$$

$$b_2 = \lambda_n(\lambda_s + \theta_n) + \theta_s \lambda_s \tag{12.44}$$

$$P_2(t) = b_4 + b_5 e^{x_2 t} - b_6 e^{x_1 t} \tag{12.45}$$

where

$$b_3 = \frac{1}{x_2 - x_1} \tag{12.46}$$

$$b_4 = \lambda_n(\lambda_s + \theta_n)/x_1 x_2 \tag{12.47}$$

$$b_5 = b_3(\lambda_n + b_4 x_1) \tag{12.48}$$

$$b_6 = b_3(\lambda_n + b_4 x_2) \tag{12.49}$$

$$P_1(t) = \theta_s b_3 \left(e^{x_2 t} - e^{x_1 t}\right) \tag{12.50}$$

$$P_3(t) = b_7 \left[(1 + b_3)\left(x_1 e^{x_2 t} - x_2 e^{x_1 t}\right)\right] \tag{12.51}$$

where

$$b_7 = (\lambda_s \theta_s)/x_1 x_2 \tag{12.52}$$

The performance reliability of the nuclear power plant worker is expressed by

$$R_{npw} = P_0(t) + P_1(t) \tag{12.53}$$

where
$R_{npw}(t)$ is the nuclear power plant worker reliability at time t.

The mean time to human error of the nuclear power plant worker is expressed by

$$MTTHE_{npw} = \int_0^\infty R_{npw}(t)dt$$
$$= (\lambda_s + \theta_s + \theta_n)/b_2 \qquad (12.54)$$

where

$MTTHE_{npw}$ is the mean time to human error of the nuclear power plant worker.

EXAMPLE 12.5

Assume that a worker is carrying out a certain task at a nuclear power plant in fluctuating environments, and his/her error rates in normal and stressful environments are 0.0001 errors/hour and 0.0005 errors/hour, respectively. The constant transition rates from stressful to normal environment and vice versa are 0.02 times/hour and 0.06 times/hour, respectively.

Calculate the mean time to human error of the nuclear power plant worker.

By substituting the specified data values into Equation 12.54, we get

$$MTTHE_{npw} = \frac{(0.0005 + 0.06 + 0.02)}{(0.0001)(0.0005 + 0.02) + (0.06)(0.0005)}$$
$$= 2511.7 \text{ hours}$$

Thus, the mean time to human error of the nuclear power plant worker is 2511.7 hours.

12.7 Model VI

Past experiences over the years indicate that human error by workers in nuclear power plants can cause not only the failure of single unit systems but also of redundant unit systems. Thus, this mathematical model represents a two-identical-units parallel system subjected to periodic preventive maintenance or other activities by workers in nuclear power plants. The system/unit can fail due to hardware failures or human errors. The state space diagram of the system is shown in Figure 12.5 and the numerals in boxes denote system states.

The following assumptions are associated with the diagram/model:

- Both system units are independent, active, and identical.
- Human errors may occur when either both system units are good or when one system unit is good.

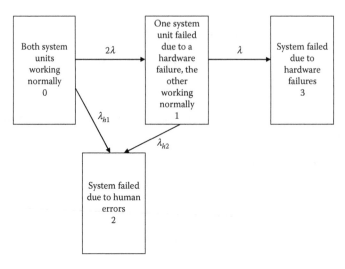

FIGURE 12.5
Two-identical-units parallel system subjected to human errors in nuclear power plants state space diagram.

- Both failure and human error rates are constant.
- All failures and human errors occur independently.

The following symbols are associated with the model/diagram:

j is the jth state of the system: $j = 0$ (both system units working normally), $j = 1$ (one system unit failed due to hardware failure, the other working normally), $j = 2$ (system failed due to human errors), $j = 3$ (system failed due to hardware failures).

$P_j(t)$ is the probability that the system is in state j at time t, for $j = 0, 1, 2, 3$.

λ_{h1} is the constant human error rate when both system units are working normally.

λ_{h2} is the constant error rate when only one unit is working normally.

λ is the unit constant failure rate.

With the aid of the Markov method, we write the following set of differential equations for Figure 12.5 diagram [2,11]:

$$\frac{dP_0(t)}{dt} + (2\lambda + \lambda_{h1})P_0(t) = 0 \tag{12.55}$$

$$\frac{dP_1(t)}{dt} + (\lambda + \lambda_{h2})P_1(t) = 2\lambda P_0(t) \tag{12.56}$$

$$\frac{dP_2(t)}{dt} = \lambda_{h1}P_0(t) + \lambda_{h2}P_1(t) \tag{12.57}$$

$$\frac{dP_3(t)}{dt} = \lambda P_1(t) \tag{12.58}$$

At time $t = 0, P_0(0) = 1, P_1(0) = 0, P_2(0) = 0,$ and $P_3(0) = 0.$

By solving Equations 12.55 through 12.58, we get

$$P_0(t) = e^{-C_1 t} \tag{12.59}$$

where

$$C_1 = 2\lambda + \lambda_{h1} \tag{12.60}$$

$$P_1(t) = D_1\left(e^{-C_1 t} - e^{-C_2 t}\right) \tag{12.61}$$

where

$$C_2 = \lambda + \lambda_{h2} \tag{12.62}$$

$$D_1 = \frac{2\lambda}{(C_2 - C_1)} \tag{12.63}$$

$$P_2(t) = D_2 - D_3 e^{-C_1 t} - D_4 e^{-C_2 t} \tag{12.64}$$

where

$$D_2 = \frac{2\lambda\lambda_{h2} + \lambda_{h1}C_2}{C_1 C_2} \tag{12.65}$$

$$D_3 = \frac{2\lambda\lambda_{h2} + \lambda_{h1}(C_2 - C_1)}{C_1(C_2 - C_1)} \tag{12.66}$$

$$D_4 = \frac{2\lambda\lambda_{h2}}{C_1(C_1 - C_2)} \tag{12.67}$$

$$P_2(t) = D_5 - D_6 e^{-C_1 t} - D_7 e^{-C_2 t} \tag{12.68}$$

where

$$D_5 = \frac{2\lambda^2}{C_1 C_2} \tag{12.69}$$

$$D_6 = \frac{2\lambda^2}{C_1(C_2 - C_1)} \tag{12.70}$$

$$D_7 = \frac{2\lambda^2}{C_2(C_1 - C_2)} \tag{12.71}$$

The system reliability is expressed by

$$R_s(t) = P_0(t) + P_1(t)$$
$$= e^{-C_1 t} + D_1\left(e^{-C_1 t} - e^{-C_2 t}\right) \tag{12.72}$$

where
$R_s(t)$ is the system reliability at time t.

The system mean time to failure is expressed by [2,4,11].

$$MTTF_s = \int_0^\infty R_s(t)dt$$
$$= \int_0^\infty \left[e^{-C_1 t} + D_1\left(e^{-C_1 t} - e^{-C_2 t}\right)\right]dt \tag{12.73}$$
$$= \frac{3\lambda + \lambda_{h2}}{(2\lambda + \lambda_{h1})(2\lambda + \lambda_{h2})}$$

where
$MTTF_s$ is the system mean time to failure.

EXAMPLE 12.6

Assume that a system in a nuclear power plant is composed of two identical and independent units in parallel. When both the units operate normally, the constant human error rate is 0.006 errors/hour, and when only one unit operates normally, the constant human error rate is 0.002 errors/hour. The constant failure rate of a unit is 0.01 failures/hour. Calculate the system mean time to failure.

By inserting the given data values into Equation 12.73, we obtain

$$MTTF_s = \frac{3(0.01) + 0.002}{[2(0.01) + 0.006][2(0.01) + 0.002]}$$
$$= 55.94$$

Thus, the system mean time to failure is 55.94 hours.

12.8 Model VII

This mathematical model represents the Model VI system subjected to repair. Thus, this model represents a system with two identical and independent units forming a parallel configuration subjected to failed unit(s) repair. The system/unit can fail due to hardware failures or human errors. The state space diagram of the system is shown in Figure 12.6 and the numerals in boxes denote system states.

The following assumptions are associated with the model:

- Failure, repair, and human error rates are constant.
- Failures and human errors occur independently.
- The total system fails due to human errors.
- Human errors may occur when either both system units are working normally or when one system unit is working normally.
- The repaired unit or system is as good as new.

The following symbols are associated with Figure 12.6 diagram:

i is the ith state of the system; $i = 0$ (both system units working normally), $i = 1$ (one system unit failed due to a hardware failure, the other working normally), $i = 2$ (system failed due to human errors), $i = 3$ (system failed due to hardware failures).

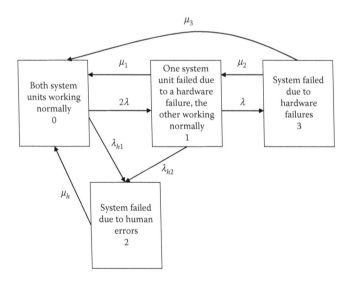

FIGURE 12.6

Two-identical-units parallel system subjected to human errors in nuclear power plants state space diagram.

$P_i(t)$ is the probability that the system is in state i at time t, for $i = 0, 1, 2, 3$.

λ is the unit constant failure rate.

λ_{h2} is the constant human error rate when only one unit is working normally.

λ_{h1} is the constant human error rate when both units are working normally.

μ_h is the system constant repair rate from state 2 to state 0.

μ_1 is the system constant repair rate from state 1 to state 0.

μ_2 is the system constant repair rate from state 3 to state 1.

μ_3 is the system constant repair rate from state 3 to state 0.

With the aid of the Markov method and Figure 12.6, we write the following set of differential equations [2,11,13]:

$$\frac{dP_0(t)}{dt} + (2\lambda + \lambda_{h1})P_0(t) = P_1(t)\mu_1 + P_3(t)\mu_3 + P_2(t)\mu_h \tag{12.74}$$

$$\frac{dP_1(t)}{dt} + (\lambda + \lambda_{h2} + \mu_1)P_1(t) = P_0(t)2\lambda + P_3(t)\mu_2 \tag{12.75}$$

$$\frac{dP_2(t)}{dt} + \mu_h P_2(t) = P_0(t)\lambda_{h1} + P_1(t)\lambda_{h2} \tag{12.76}$$

$$\frac{dP_3(t)}{dt} + (\mu_2 + \mu_3)P_3(t) = P_1(t)\lambda \tag{12.77}$$

At time $t = 0, P_0(0) = 1, P_1(0) = 0, P_2(0) = 0,$ and $P_3(0) = 0$.

By solving Equations 12.74 through 12.77, we get the following steady-state probability equations [2,11]:

$$P_0 = \left[1 + A_1 + 2\lambda^2 A + \frac{1}{\mu_h}(\lambda_{h1} + A_1\lambda_{h2})\right]^{-1} \tag{12.78}$$

where

$$A = \left[(\mu_3 + \mu_2)(\lambda + \lambda_{h2} + \mu_1) - \lambda\mu_2\right]^{-1} \tag{12.79}$$

$$A_1 = 2\lambda(1 + \lambda\mu_2 A)/(\lambda + \lambda_{h2} + \mu_1) \tag{12.80}$$

$$P_1 = P_0 A_1 \tag{12.81}$$

$$P_2 = P_0(\lambda_{h1} + A_1\lambda_{h2})/\mu_h \tag{12.82}$$

$$P_3 = P_0 2\lambda^2 A \tag{12.83}$$

where
P_0, P_1, P_2, and, P_3 are the steady-state probabilities of the system being in states 0, 1, 2, and 3, respectively.

The system steady-state availability is expressed by

$$AV_s = P_0 + P_1 \tag{12.84}$$

where
AV_s is the system steady-state availability.

Additional information on this mathematical model is available in References 2 and 13.

12.9 Model VIII

This mathematical model represents a two-identical redundant active units nuclear power plant system subjected to human errors. Each of these units can malfunction either due to the occurrence of a hardware failure or a human error. At least one of these units must work normally for the successful operation of the nuclear power plant system.

The state space diagram of the system is shown in Figure 12.7. The numerals in the boxes and circles denote system states.

The following assumptions are associated with the diagram/model:

- Each system unit can fail due to a hardware failure or human error.
- Both the system units operate simultaneously and are identical.
- Hardware failure and human error rates are constant.
- Hardware failures and human errors occur independently.
- Failures of each unit can be categorized under two classifications: failures due to hardware problems and failures due to human errors.

The following symbols are associated with the diagram/model:

i is the ith state of the nuclear power plant system: $i = 0$ (both units of the nuclear power plant system working normally), $i = 1$ (one unit failed due to a hardware failure, the other working normally), $i = 2$ (one unit failed due to a human error, the other working normally), $i = 3$ (both units failed due to hardware failures), and $i = 4$ (both units failed due to human errors).

$P_i(t)$ is the probability that the nuclear power plant system is in state i at time t, for $i = 0, 1, 2, 3, 4$.

λ_1 is the constant hardware failure rate of a unit.

λ_2 is the constant human error rate of a unit.

By using the Markov method, we write the following set of differential equations for the Figure 12.7 diagram [2]:

$$\frac{dP_0(t)}{dt} + (2\lambda_1 + 2\lambda_2)P_0(t) = 0 \tag{12.85}$$

$$\frac{dP_1(t)}{dt} + (\lambda_1 + \lambda_2)P_1(t) = 2\lambda_1 P_0(t) \tag{12.86}$$

$$\frac{dP_2(t)}{dt} + (\lambda_1 + \lambda_2)P_2(t) = 2\lambda_2 P_0(t) \tag{12.87}$$

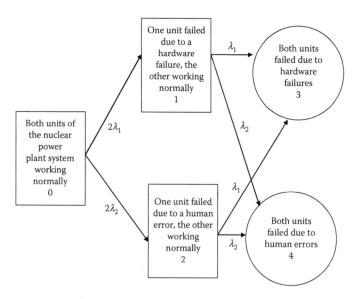

FIGURE 12.7
Two-redundant unit nuclear power plant system state space diagram.

$$\frac{dP_3(t)}{dt} = \lambda_1 P_1(t) + \lambda_1 P_2(t) \tag{12.88}$$

$$\frac{dP_4(t)}{dt} = \lambda_2 P_1(t) + \lambda_2 P_2(t) \tag{12.89}$$

At time $t = 0$, $P_0(0) = 1$, and $P_1(0) = P_2(0) = P_3(0) = P_4(0) = 0$.

By solving Equations 12.85 through 12.89, we get the following equations for the nuclear power plant system state probabilities:

$$P_0(t) = e^{-2(\lambda_1 + \lambda_2)t} \tag{12.90}$$

$$P_1(t) = \frac{2\lambda_1}{(\lambda_1 + \lambda_2)} \left[e^{-(\lambda_1 + \lambda_2)t} - e^{-2(\lambda_1 + \lambda_2)t} \right] \tag{12.91}$$

$$P_2(t) = \frac{2\lambda_2}{(\lambda_1 + \lambda_2)} \left[e^{-(\lambda_1 + \lambda_2)t} - e^{-2(\lambda_1 + \lambda_2)t} \right] \tag{12.92}$$

$$P_3(t) = \frac{\lambda_1}{(\lambda_1 + \lambda_2)} \left[1 - e^{-(\lambda_1 + \lambda_2)t} \right]^2 \tag{12.93}$$

$$P_4(t) = \frac{\lambda_2}{(\lambda_1 + \lambda_2)} \left[1 - e^{-(\lambda_1 + \lambda_2)t} \right]^2 \tag{12.94}$$

The nuclear power plant system reliability is given by

$$R_{ns}(t) = P_0(t) + P_1(t) + P_2(t)$$
$$= 1 - \left[1 - e^{-(\lambda_1 + \lambda_2)t} \right]^2 \tag{12.95}$$

where
$R_{ns}(t)$ is the nuclear power plant system reliability at time t.

By integrating Equation 12.94 over the time interval $[0,\infty]$, we get [4]:

$$MTTF_{ns} = \int_0^\infty R_{ns}(t)dt$$

$$= \frac{3}{2(\lambda_1 + \lambda_2)} \tag{12.96}$$

where
$MTTF_{ns}$ is the mean time to failure of a two-redundant active unit nuclear power plant system with human error.

EXAMPLE 12.7

Assume that a nuclear power plant system is composed of two active, identical, and independent units. At least one unit must operate normally for the system's successful operation. Each unit can fail either due to a hardware failure or due to a human error. Constant hardware failure and human error rates are 0.004 failures/hour and 0.0005 errors/hour, respectively.

Calculate the nuclear power plant system mean time to failure.

By inserting the given data values into Equation 12.96, we get

$$MTTF_{ns} = \frac{3}{2(0.004 + 0.0005)}$$
$$= 333.3 \text{ hours}$$

Thus, the nuclear power plant system mean time to failure is 333.3 hours.

12.10 Problems

1. Assume that a nuclear power plant system can fail safely or unsafely and its constant failure rates are 0.0009 failures per hour and 0.0001 failures per hour, respectively. Calculate the probability of the nuclear power plant system failing unsafely during a 100-hour mission and its mean time to failure.

2. Prove Equations 12.16 through 12.18 by using Equations 12.13 through 12.15 and the given initial conditions.

3. A worker in a nuclear power plant is performing a maintenance task and his/her error rate is 0.006 errors/hour. Calculate the worker's mean time to human error and reliability during a seven-hour work period.

4. Prove Equation 12.26 by using Equations 12.19 and 12.20.

5. Assume that a nuclear power plant system can fail either due to hardware failures or human errors made by nuclear power plant workers. The system constant hardware failure and human error rates are 0.008 failures/hour and 0.001 errors/hour, respectively. Calculate the probability that the system will fail due to a human error made by nuclear power plant workers during a 24-hour mission.

6. Prove Equations 12.40, 12.45, 12.50, and 12.51 by using Equations 12.36 through 12.39 and the given initial conditions.

7. Prove Equations 12.73 by using Equation 12.72.

8. Assume that a system in a nuclear power plant is composed of two identical and independent units in parallel. When both the units

operate normally, the constant human error rate is 0.004 errors/hour, and when only one unit operates normally, the constant human error rate is 0.001 errors/hour. The constant failure rate of a unit is 0.03 failures/hour. Calculate the system mean time to failure.

9. Prove Equations 12.78, 12.81, 12.82, and 12.83 by using Equations 12.74 through 12.77 and the given initial conditions.

10. Prove Equation 12.96 by using Equations 12.90 through 12.92.

References

1. Regulinski, T.L., Askren, W.B., Stochastic modeling of human performing effectiveness functions, *Proceedings of the Annual Reliability and Maintainability Symposium*, 1972, pp. 407–416.
2. Dhillon, B.S., *Human Reliability: With Human Factors*, Pergamon Press, New York, 1986.
3. Dhillon, B.S., *Engineering Safety: Fundamentals, Techniques, and Applications*, World Scientific Publishing, River Edge, New Jersey, 2003.
4. Dhillon, B.S., *Design Reliability: Fundamentals and Applications*, CRC Press, Boca Raton, Florida, 1999.
5. Dhillon, B.S., *Safety and Human Error in Engineering Systems*, CRC Press, Boca Raton, Florida, 2013.
6. Regulinski, T.L., Askren, W.B., Mathematical modeling of human performance reliability, *Proceedings of the Annual Symposium on Reliability*, 1969, pp. 5–11.
7. Dhillon, B.S., The Analysis of the Reliability of Multi-State Device Networks, Ph.D. Dissertation, 1975. Available from the National Library of Canada, Ottawa, Canada.
8. Dhillon, B.S., *Mine Safety: A Modern Approach*, Springer, London, 2010.
9. Shooman, M.L., *Probabilistic Reliability: An Engineering Approach*, McGraw-Hill, New York, 1968.
10. Askren, W.B., Regulinsi, T.L., Quantifying human performance for reliability analysis of systems, *Human Factors*, Vol. 11, 1969, pp. 393–396.
11. Dhillon, B.S., *Human Reliability, Error, and Human Factors in Engineering Maintenance: with Reference to Aviation and Power Generation*, CRC Press, Boca Raton, Florida, 2009.
12. Dhillon, B.S., Stochastic models for predicting human reliability, *Microelectronics and Reliability*, Vol. 25, 1985, pp. 729–752.
13. Dhillon, B.S., Rayapati, S.N., Analysis of redundant systems with human errors, *Proceedings of the Annual Reliability and Maintainability Symposium*, 1985, pp. 315–321.

Appendix: Bibliography—Literature on Safety, Reliability, Human Factors, and Human Error in Nuclear Power Plants

A.1 Introduction

Over the years, a large number of publications on safety, reliability, human factors, and human error in nuclear power plants have appeared in the form of journal articles, conference proceedings articles, technical reports, and so on. This appendix presents an extensive list of selective publications related, directly or indirectly, to safety, reliability, human factors, and human error in nuclear power plants. The period covered by the listing is from 1960 to 2015. The main objective of this listing is to provide readers with sources for obtaining additional information on safety, reliability, human factors, and human error in nuclear power plants.

Publications

1. Abu-Khader, M.M., Recent advances in nuclear power: A review, *Progress in Nuclear Energy*, Vol. 51, No. 2, 2009, pp. 225–235.
2. Ahmed, I. et al., A computer based living probabilistic safety assessment (LPSA) method for nuclear power plants, *Nuclear Engineering and Design*, Vol. 265, 2013, pp. 765–771.
3. Aldemir, T. et al., Dynamic reliability modeling of digital instrumentation and control systems in nuclear power plants, *Proceedings of the 6th American Nuclear Society International Topical Meeting on Nuclear Plant Instrumentation, Control, and Human-Machine Interface Technologies*, 2009, pp. 1197–1206.
4. Alder, G.C., Todd, F.J., Nuclear power plant thermal performance improvement, *Proceedings of the ASME Power Conference*, 2008, pp. 753–761.
5. Ali, E.A.M., Muzzammil, M.H.S.M., Safety improvement of nuclear power reactor using soft computing techniques, *Proceedings of the International Conference on Energy Efficient Technologies for Sustainability*, 2013, pp. 949–954.
6. Alzbutas, R., Voronov, R., Reliability and safety analysis for systems of fusion device, *Fusion Engineering and Design*, Vol. 94, 2015, pp. 31–41.

7. Ammirato, V.J., Mui, M.C., Realistic RAM goals on the Indian Point 3 Nuclear Power Plant distribution system upgrade conceptual engineering study, *Proceedings of the 13th INTER-RAM: International Reliability, Availability, Maintainability Conference for the Electric Power Industry*, 1986, pp. 198–201.

8. Ananda, M.M.A., Reliability modeling of engineered barrier systems for nuclear waste: A conditional approach, *Microelectronics and Reliability*, Vol. 34, No. 7, 1994, pp. 1221–1225.

9. Annick, C., An EDF perspective on human factors, *Proceedings of the IEEE Conference on Human Factors and Power Plants*, 1988, pp. 65–67.

10. Aoyagi, S. et al., A discussion system for knowledge sharing and collaborative analysis of incidents in nuclear power plants, *Proceedings of the Third International Online Communities and Social Computing Conference*, 2009, pp. 3–12.

11. Arndt, S.A., Integrating software reliability concepts into risk and reliability modeling of digital instrumentation and control systems used in nuclear power plants, *Proceedings of the 5th International Topical Meeting on Nuclear Plant Instrumentation Controls, and Human Machine Interface Technology*, 2006, pp. 849–851.

12. ArunBabu, P., Senthil Kumar, C., Murali, N., A hybrid approach to quantify software reliability in nuclear safety systems, *Analysis of Nuclear Energy*, Vol. 50, 2012, pp. 133–141.

13. Augutis, Y., Uspuras, E., Alzbutas, R., Matuzas, V., Reliability of the coolant flow rate measuring system at the Ignalina Nuclear Power Plant, *Atomic Energy*, Vol. 96, No. 6, 2004, pp. 403–406.

14. Austin, R.C., 500 MW unit control center: Operations and maintenance oriented, *Instrumentation in Power Industry*, Vol. 11, 1988, pp. 56–66.

15. Ayres, T.J., Gross, M.M., Exploring power plant data resources for organizational epidemiology, *Proceedings of the IEEE Conference on Human Factors and Power Plants*, 2002, pp. 10–16.

16. Baley-Downes, S., Human reliability program-components and effects, *Nuclear Materials Management*, Vol. 15, 1986, pp. 661–665.

17. Banks, W.W., Blackman, H.S., Curtis, J.N., A human factors evaluation of nuclear power plant control room annunciators, *Proceedings of the Annual Control Engineering Conference*, 1984, pp. 323–326.

18. Baranenko, V.I. et al., Reliability of high-pressure recovery system in the power-generating units of nuclear power plants with VVER-1000 and 440 nuclear reactors, *Atomic Energy*, Vol. 78, No. 2, 1995, pp. 129–135.

19. Baranowsky, P.W., O'Reilly, P.D., Rasmuson, D.M., A reliability and risk-based approach to analyzing operational data at U.S. nuclear power plants, *Proceedings of the International Meeting: PSA/PRA and Severe Accidents*, 1994, pp. 189–196.

20. Basu, S., Kee, F.J., Reliability study of reactor control system for a CANDU-PHW nuclear unit, *Proceedings of the IEEE Power Generation Conference*, 1977, pp. 111–113.

21. Basu, S., Zemdegs, R., Method of reliability analysis of control systems for nuclear power plants, *Microelectronics and Reliability*, Vol. 17, No. 1, 1978, pp. 105–116.

22. Baur, P., Strategies for eliminating human error in the control room, *Power*, Vol. 27, No. 5, 1983, pp. 21–29.

23. Beltracchi, L., Iconic displays, ranking cycles and human factors for control rooms of nuclear power plants, *IEEE Transactions on Nuclear Science*, Vol. 30, No. 3, 1983, pp. 856–1861.

24. Berg, R.M., Ruether, J.C., Wray, L., Wilkinson, C.D., Methodology for evaluation of maintainability and obsolescence of nuclear plant I&C systems, *Proceedings of the American Power Conference*, 1994, pp. 1256–1261.
25. Beveridge, R.L., Applied human reliability: One utility's experience, *Proceedings of the IEEE Conference on Human Factors and Nuclear Safety*, 1985, pp. 316–320.
26. Binkley, D.M., Day, M.F., Hardy, J.D., Rizzie, J.W., Signal isolation of safety-class circuits for nuclear power plants, *Proceedings of the American Power Conference*, 1982, pp. 553–558.
27. Blanchard, D. et al., Power plant generation risk assessment (GRA), *Proceedings of the American Nuclear Society International Topical Meeting on Probabilistic Safety Analysis*, 2005, pp. 915–926.
28. Blot, M., Laviron, A., Reliability analysis with the simulator S. ESCAF of a very complex sequential system: The electrical power supply system of a nuclear reactor, *Reliability Engineering & System Safety*, Vol. 21, No. 2, 1988, pp. 91–105.
29. Bongarra, J.P., Certifying advanced plants: A U.S. NRC human factors perspective, *Proceedings of the IEEE Conference on Human Factors and Power Plants*, 1997, pp. 6.18–6.24.
30. Boring, R.L. et al., A taxonomy and database for capturing human reliability and human performance data, *Proceedings of the Human Factors and Ergonomics Society Annual Meeting*, 2006, pp. 2217–2221.
31. Boring, R.L., Gertman, D.I., Joe, J.C., Marbie, J.L., Human reliability analysis in the U.S. nuclear power industry: A comparison of atomistic and holistic methods, *Proceedings of the Human Factors and Ergonomics Society Annual Meeting*, 2005, pp. 1814–1819.
32. Boring, R.L. et al., Human factors and the nuclear renaissance, *Proceedings of the Human Factors and Ergonomics Society Conference*, 2008, pp. 763–767.
33. Boring, R.L., Griffith, C.D., Joe, J.C., The measure of human error: Direct and indirect performance shaping factors, *Proceedings of the IEEE Conference on Human Factors and Power Plants*, 2007, pp. 170–176.
34. Boring, R.L., Oxstrand, J., Hildebrandt, M., Human reliability analysis for control room upgrades, *Proceedings of the Human Factors and Ergonomics Society Conference*, 2009, pp. 1584–1588.
35. Boring, R.L., Modeling human reliability analysis using MIDAS, *Proceedings of the International Workshop on Future Control Station Designs and Human Performance Issues in Nuclear Power Plants*, 2009, pp. 89–92.
36. Boring, R.L., Modeling human reliability analysis using MIDAS, *Proceedings of the International Topical Meeting on Nuclear Plant Instrumentation Control, and Human Machine Interface Technology*, 2006, pp. 1270–1274.
37. Boring, R.L., Using nuclear power plant training simulators for operator performance and human reliability research, *Proceedings of the American Nuclear Society International Topical Meeting on Nuclear Power Plant Instrumentation Controls, and Human Machine Interface Technology*, 2009, pp. 1–12.
38. Bransby, M., The human contribution to safety: Designing alarm systems for reliable operator performance, *Measurement and Control*, Vol. 32, No. 7, 1999, pp. 209–213.
39. Brooks, A.C., The application of availability analysis to nuclear power plants, *Reliability Engineering*, Vol. 9, No. 3, 1984, pp. 127–131.

40. Bucci, P. et al., A benchmark system for the assessment of reliability modeling methods for digital instrumentation and control systems in nuclear plants, *Proceedings of the 5th International Topical Meeting on Nuclear Plant Instrumentation Controls, and Human Machine Interface Technology*, 2006, pp. 277–283.

41. Buende, R., Reliability, availability, and quality assurance considerations for fusion components, *Fusion Engineering and Design*, Vol. 29, 1995, pp. 262–285.

42. Bukrinskii, A.M., Shviryaev, Y.V., Derzhinskii, F.E., Requirements for reliability of nuclear power plant (NPP) safety systems, *Soviet Power Engineering*, Vol. 10, No. 3, 1981, pp. 323–330.

43. Cacciabue, P.C., Evaluation of human factors and man-machine problems in the safety of nuclear power plants, *Nuclear Engineering and Design*, Vol. 109, No. 3, 1988, pp. 417–431.

44. Carnino, A., Human reliability, *Nuclear Engineering and Design*, Vol. 90, No. 3, 1985, pp. 365–369.

45. Carnino, A., An EDF perspective on human factors, *Proceedings of the IEEE Conference on Human Factors and Power Plants*, 1988, pp. 65–68.

46. Carpignano, A., Piccini, M., Cognitive theories and engineering approaches for safety assessment and design of automated systems: A case study of power plant, *Cognition Technology & Work*, Vol. 11, No. 1, 1999, pp. 47–61.

47. Carvalho, P.V.R. et al., Human factors approach for evaluation and redesign of human-system interfaces of a nuclear power plant simulator, *Displays*, Vol. 29, No. 3, 2008, pp. 273–284.

48. Cepin, M., Prosek, A., Update of the human reliability analysis for a nuclear power plant, *Proceedings of the International Conference on Nuclear Energy for New Europe*, 2006, pp. 706.1–706.8.

49. Chang, S.H. et al., Development of an advanced human-machine interface for next generation nuclear power plants, *Reliability Engineering & System Safety*, Vol. 64, No. 1, 1999, pp. 109–126.

50. Chang, Y.H.J., Coyne, K., Mosleh, A., A nuclear plant accident diagnosis method to support prediction of errors of commission, *Proceedings of the 5th International Topical Meeting on Nuclear Plant Instrumentation Controls, and Human Machine Interface Technology*, 2006, pp. 1261–1269.

51. Chaojun, L. et al., Analysis on the reliability assurance program of advanced nuclear power plant operating, *Proceedings of the 10th International Conference on Reliability, Maintainability and Safety*, 2014, pp. 1008–1011.

52. Chiba, M., Safety and reliability: A case study of operating Ikata Nuclear Power Plant, Japan, *Journal of Engineering and Technology Management*, Vol. 7, No. 3–4, 1991, pp. 267–278.

53. Chiramal, M., Application of commercial-grade digital equipment in nuclear power plant safety systems, *Proceedings of the IEEE Symposium on Reliable Distributed Systems*, 2001, pp. 176–178.

54. Cho, N.Z., Jung, W.S., Semi-Markov reliability analysis of three test/repair policies for standby safety systems in a nuclear power plant, *Reliability Engineering & System Safety*, Vol. 31, No. 1, 1991, pp. 1–30.

55. Cho, S., Jiang, J., Analysis of surveillance test interval by Markov process for SDS1 in CANDU nuclear power plants, *Reliability Engineering and System Safety*, Vol. 93, No. 1, 2008, pp. 1–13.

56. Choi, S.S. et al., Development strategies of an intelligent human-machine interface for next generation nuclear power plants, *IEEE Transactions on Nuclear Science*, Vol. 43, No. 3, 1996, pp. 2096–2114.

57. Chung, H.Y., Lee, Y.H., Strategy for establishing integrated I&C reliability of operating nuclear power plants in Korea, *Proceedings of the International Conference on Advances in Nuclear Power Plants*, 2008, pp. 899–903.

58. Clarke, L.R., Allamby, S.P., People in power system control in the next century, *Proceedings of the International Conference on Human Interfaces in Control Room, Cockpits, and Command Centres*, 1999, pp. 434–439.

59. Cockrell, J.L., Magee, J.H., Underkoffler, V.S., Nuclear protective system design for reliability, *IEEE Transactions on Communications and Electronics*, Vol. 67, 1963, pp. 388–402.

60. Cummings, G.E., An application of system reliability analysis for the study of reactor seismic safety, *Nuclear Engineering and Design*, Vol. 71, No. 3, 1982, pp. 341–344.

61. Dalrymple, D.G., Nuclear plant life assurance scoping study nuclear & conventional piping, *Proceedings of the 2nd International Conference on CANDU Maintenance*, 1992, pp. 159–171.

62. Daly, S., Orme, S., The reliability of the Sizewell "B" reactor protection system, *Proceedings of the International Conference on Electrical and Control Aspects of the Sizewell B PWR*, 1992, pp. 208–214.

63. Damsin, M., Periodic safety reassessment of Belgian nuclear power plants, *Proceedings of the International Symposium on Advances in the Operational Safety of Nuclear Power Plants*, 1996, pp. 491–499.

64. Danchak, M.M., CRT displays for power plants, *Instrumentation Technology*, Vol. 23, No. 10, 1976, pp. 29–36.

65. Daniel, S.E., Heising, C.D., Impact of the organization on human reliability at Ft. Calhoun nuclear plant, *Transactions of the American Nuclear Society*, Vol. 88, 2003, pp. 888–889.

66. Dang, V.N. et al., An empirical study of HRA methods—overall design and issues, *Proceedings the IEEE Conference on Human Factors and Power Plants*, 2007, pp. 243–247.

67. Dang, V., Huang, Y., Siu, N., Carroll, J., Analyzing cognitive errors using a dynamic crew-simulation model, *Proceedings of the IEEE Conference on Human Factors and Power Plants*, 1992, pp. 520–526.

68. Daniels, R.W., Formula for improved plant maintainability must include human factors, *Proceedings of the IEEE Conference on Human Factors and Nuclear Safety*, 1985, pp. 242–244.

69. Danielsson, H., Preventive maintenance at the Forsmark Nuclear Power Plant, *Proceedings of the International Symposium on Advances in Nuclear Power Plant Availability, Maintainability and Operation*, 1985, pp. 217–223.

70. Denning, R.S., Durst, B.M., Fletcher, N., Safety parameter display systems for Soviet-designed nuclear power plants, *Proceedings of the International Topical Meeting on Safety of Operating Reactors*, 1998, pp. 33–39.

71. Der Kiureghian, A., Moghtaderi-Zadeh, M., An integrated approach to the reliability of engineering systems, *Nuclear Engineering and Design*, Vol. 71, No. 3, 1982, pp. 349–354.

72. Dhillon, B.S., Bibliography of literature on nuclear system reliability, *Microelectronics Reliability*, Vol. 23, No. 6, 1983, pp. 1143–1161.

73. Dougherty, E.M., Fragola, J.R., Foundations for a time reliability correlation system to quantify human reliability, *Proceedings of the IEEE Conference on Human Factors and Power Plants*, 1988, pp. 268–278.

74. Duffey, R.B., Saull, J.W., The probability and management of human error, *Proceedings of the International Conference on Nuclear Energy*, 2004, pp. 133–137.

75. Ewing, T.F., Anastasia, L.J., Haskell, R.M., Nuclear plant reliability prediction based on historical maintenance data, *Nuclear Plant Journal*, Vol. 5, No. 3, 1987, pp. 56–64.

76. Farinha, T. et al., New approaches on predictive maintenance based on an environmental perspective, the cases of wind generators and diesel engines, *Proceedings of the 12th WSEAS International Conference on Circuits*, 2008, pp. 420–428.

77. Felkel, L., STAR-GENERIS: A generic software package enhancing reliability and availability of nuclear power plants, *Proceedings of the 13th INTER-RAM: International Reliability, Availability, Maintainability Conference for the Electric Power Industry*, 1986, pp. 192–197.

78. Fenton, E.F. et al., Development of a unit control system concept for Lakeview Generating Station, *Proceedings of Instrumentation, Control, and Automation in the Power Industry*, Vol. 33, 1990, pp. 245–250.

79. Ferguson, K., Technology success: Integration of power plant reliability and effective maintenance, *Proceedings of the International Conference on Advances in Nuclear Power Plants*, 2008, pp. 938–944.

80. Fiedlander, M.A., Evans, S.A., Influence of organizational culture on human error, *Proceedings of the IEEE Conference on Human Factors and Power Plants*, 1997, pp. 19–22.

81. Fink, B., Hill, D., Naser, J., Planning for control room modernization, *Proceedings of the American Nuclear Society International Topical Meeting on Nuclear Plant Instrumentation, Control, and Human Machine Interface Technology*, 2004, pp. 1475–483.

82. Fischer, R.E., Specialist maintenance techniques for the nuclear industry, *Nuclear Engineer*, Vol. 33, No. 3, 1992, pp. 74–80.

83. Fleger, S.A., McWilliams, M.R., Control room human factors assessment at Bulgarian power plants, *Proceedings of the Human Factors and Ergonomics Society Meeting*, 1995, pp. 1043–1047.

84. Floyd II, H.L., Industrial electric power system operation, Part I: Improving system reliability, *Proceedings of the IEEE Industrial and Commercial Power Systems Technical Conference*, 1984, pp. 153–158.

85. Floyd II, H.L., Reducing human errors in industrial electric power system operation, Part I: Improving system reliability, *IEEE Transactions on Industry Applications*, Vol. 22, No. 3, 1986, pp. 420–424.

86. Foerster, F., Slanina, S., A new concept for filtration units for trapping radioactive aerosols and iodine in the ventilation systems of nuclear power plants with WWER reactors, *Proceedings of the International Symposium on Advances in Nuclear Power Plant Availability, Maintainability and Operation*, 1985, pp. 95–101.

87. Forell, B.E., Einarsson, S., Röwekamp, M., Berg, H.P., Updated technical reliability data for fire protection systems and components at a German nuclear power plant, *Proceedings of the 11th International Probabilistic Safety Assessment and Management Conference and the Annual European Safety and Reliability Conference*, 2012, pp. 3783–3794.

88. Forzano, P., Castagna, P., Procedures, quality, standards, and the role of human factors and computerized tools in power plant control, *Proceedings of the IEEE Conference on Human Factors and Power Plants*, 1997, pp. 7–12.

89. Fragola, J.R., Shooman, M.L., Experience bounds on nuclear plant probabilistic safety assessment, *Proceedings of the Annual Reliability and Maintainability Symposium*, 1992, pp. 157–165.

90. Fredricks, P., Humanizing the power plant control room, *Instrumentation in the Power Industry*, Vol. 23, 1980, pp. 149–153.

91. Fuchs, S., Hale, K.S., Axelsson, P., Augmented cognition can increase human performance in the control room, *Proceedings of the IEEE Conference on Human Factors and Power Plants*, 2007, pp. 128–132.

92. Furuhama, Y., Furuta, K., Kondo, S., Identification of causes of human errors in support of the development of intelligent computer-assisted instruction systems for plant operator training, *Reliability Engineering and System Safety*, Vol. 47, No. 2, 1995, pp. 75–84.

93. Fussell, J.B., Nuclear power system reliability: A historical perspective, *IEEE Transactions on Reliability*, Vol. 33, No. 1, 1984, pp. 41–77.

94. Gaddy, C.D., Taylor, J.C., Fray, R.R., Divakaruni, S.M., Human factors considerations in digital control room upgrades, *Proceedings of the American Power Conference*, 1991, pp. 256–258.

95. García-Herrero, S. et al., Bayesian network analysis of safety culture and organizational culture in a nuclear power plant, *Safety Science*, Vol. 53, 2013, pp. 82–95.

96. Gjorgiev, B., Volkanovski, A., Renewable sources impact on power system reliability and nuclear safety, *Proceedings of the European Safety and Reliability Conference*, 2015, pp. 57–63.

97. Gofuku, A., Niwa, Y., Extraction of mental models of operators on safety functions of nuclear power plants, *Proceedings of the 8th IFAC/IFIP/IFORS/IEA Symposium on Analysis, Design, and Evaluation of Human-Machine Systems*, 2001, pp. 541–546.

98. Yang, G.X., Chi, G.X., Jiang, J.Z., A case study for the design of visual environment of a control room in a power plant, *Lighting Research and Technology*, Vol. 17, No. 2, 1985, pp. 84–88.

99. Gradin, L.P., Utilizing reliability concepts in the development of IEEE recommended good practices for nuclear plant maintenance, *Proceedings of the 13th INTER-RAM: International Reliability, Availability, Maintainability Conference for the Electric Power Industry*, 1986, pp. 168–172.

100. Groth, K., Mosleh, A., Data-driven modelling of dependencies among influencing factors in human-machine interactions, *Proceedings of the ANS/PSA Topical Meeting on Challenges to PSA During the Nuclear Renaissance*, 2008, pp. 916–926.

101. Gupta, J.P. et al., Reliability trending for nuclear power plant relicensing, *Proceedings of the 2nd International Conference on Reliability, Safety and Hazard—Risk-Based Technologies and Physics-of-Failure Methods*, 2010, pp. 391–394.

102. Guttmann, H.E., Human reliability data: The problem and some approaches, *Transactions of the American Nuclear Society*, Vol. 41, No. 1, 1982, pp. 1–25.

103. Ha, J.S., Seong, P.H., Lee, M.S., Hong, J.H., Development of human performance measures for human factors validation in the advanced MCR of APR-1400, *IEEE Transactions on Nuclear Science*, Vol. 54, No. 6, 2007, pp. 2687–2699.

104. Hagen, E.W., Quantification of human error associated with instrumentation and control system components, *Nuclear Safety*, Vol. 23, No. 6, 1982, pp. 665–668.
105. Hallbert, B.P., The evaluation of human reliability in process systems analysis, *Proceedings of the IEEE Conference on Human Factors and Power Plants*, 1992, pp. 442–447.
106. Halminen, J., Nevalainen, R., Qualification of safety-critical systems in TVO nuclear power plants, *Software Process: Improvement and Practice*, Vol. 12, No. 6, 2007, pp. 559–667.
107. Hannaman, G.W., The role of framework, models, data, and judgement in human reliability, *Nuclear Engineering and Design*, Vol. 93, 1986, pp. 295–301.
108. Hart, R.S., Constructability and maintainability of nuclear power plant, *Proceedings of the Symposium on Advances in Nuclear Power Plant Availability, Maintainability and Operation*, 1985, pp. 67–80.
109. Hashemian, H.M., Fain, R.E., Reducing outage time through automated tests to meet technical specification requirements nuclear power plants, *Proceedings of the American Nuclear Society International Topical Meeting on Nuclear Plant Instrumentation, Control and Human-Machine Interface Technologies*, 1996, pp. 1289–1296.
110. Hattori, T., Maintenance management of nuclear power plants in Japan— Present situation of preventive maintenance, *Proceedings of the International Nuclear Power Plant Aging Symposium*, 1989, pp. 291–296.
111. Heckle, W.L., Bolian, T.W., Plant modernization with digital reactor protection system: Safety system upgrades at US nuclear power stations, *Proceedings of the International Congress on Advances in Nuclear Power Plants*, 2006, pp. 884–890.
112. Hedden, O.F., Cowfer, C.D., Progress toward regulatory acceptance of risk-informed inspection programs for nuclear power plants, *Proceedings of the SPIE—The International Society for Optical Engineering Conference*, 1996, pp. 41–45.
113. Heo, G., Park, J., Framework of quantifying human error effects in testing and maintenance, *Proceedings of the American Nuclear Society International Topical Meeting on Nuclear Plant Instrumentation, Control, and Human-Machine Interface Technology*, 2009, pp. 2083–2092.
114. Heo, G., Park, J., A framework for evaluating the effects of maintenance-related human errors in nuclear power plants, *Reliability Engineering and System Safety*, Vol. 95, No. 7, 2010, pp. 797–805.
115. Heuertz, S.W., Herrin, J.L., Validation of existing nuclear station instrumentation and electric procedures to reduce human errors: One utility's perspective, *IEEE Transactions on Energy Conversions*, Vol. 1, No. 4, 1986, pp. 33–34.
116. Hill, D. et al., Integration of human factors engineering into the design change process, *Proceedings of the American Nuclear Society International Topical Meeting on Nuclear Plant Instrumentation, Control, and Human-Machine Interface Technology*, 2004, pp. 1485–1494.
117. Hopkins, C.O. et al., Critical human-factors issue in nuclear power regulation and a recommended comprehensive long-range plan, *Proceedings of the Human Factors Society Annual Meeting*, 1982, pp. 692–697.
118. Howey, G.R., Human factors in Ontario hydro's nuclear power stations, *IEEE Transactions on Nuclear Science*, Vol. 28, No. 1, 1981, pp. 968–971.
119. Humphreys, M., Reliability of nuclear-power-station protective systems, *Electronics and Power*, Vol. 28, No. 7–8, 1982, pp. 511–514.

120. Hunt, R.M., Rouse, W.B., Problem-solving skills of maintenance trainees in diagnosing faults in stimulated powerplants, *Human Factors*, Vol. 23, No. 3, 1981, pp. 317–328.
121. Hvelplund, R., Control room for high-efficiency power plant, *Proceedings of the International Conference on Human Interfaces in Control Room, Cockpits and Command Centres*, 1999, pp. 101–117.
122. Jacobs, R. et al., Organizational processes and nuclear power plant safety: Research summary, *Proceedings of the IEEE Fifth Conference on Human Factors and Power Plants*, 1992, pp. 394–398.
123. Jaenkaelae, K.E., Vaurio, J.K., Vuorio, U.M., Plant-specific reliability and human data analysis for safety assessment, *Proceedings of the IAEA Conference on Nuclear Power Performance and Safety*, 1988, pp. 135–151.
124. Jervis, M.W., Nuclear reactor safety systems, *Journal of the Institution of Electrical Engineers*, Vol. 9, No. 104, 1963, pp. 351–353.
125. Jiang, J.J. et al., Association rules analysis of human factor events based on statistics method in digital nuclear power plant, *Safety Science*, Vol. 49, No. 6, 2011, pp. 946–950.
126. Jiri, S., Reliability assessment of diversity in digital I&C systems at nuclear power plants, *Proceedings of the 7th International Topical Meeting on Nuclear Plant Instrumentation, Control, and Human-Machine Interface Technologies*, 2010, pp. 1272–1281.
127. Johnson, R.D., Wise, M.J., Nuclear plant reliability data program, *Proceedings of the Annual Reliability and Maintainability Symposium*, 1975, pp. 143–148.
128. Johnson, W.B., Rouse, W.B., Analysis and classification of human errors in troubleshooting live aircraft power plants, *IEEE Transactions on Systems, Man and Cybernetics*, Vol. 12, No. 3, 1982, pp. 389–393.
129. Jou, Y.T. et al., The research on extracting the information of human errors in the main control room of nuclear power plants by using performance evaluation matrix, *Safety Science*, Vol. 49, No. 2, 2011, pp. 236–242.
130. Jou, Y.T. et al., The implementation of a human factors engineering checklist for human-system interfaces upgrade in nuclear power plants, *Safety Science*, Vol. 47, No. 7, 2009, pp. 1016–1025.
131. Jufang, S., Human errors in nuclear power plants, *Nuclear Power Engineering*, Vol. 14, No. 4, 1993, pp. 306–309.
132. Jung, W. et al., Analysis of an operator's performance time and its application to a human reliability analysis in nuclear power plants, *IEEE Transactions on Nuclear Science*, Vol. 54, No. 4, 2007, pp. 1801–1811.
133. Jung, W., Park, J., Jaewhan, K., Performance time evaluation for human reliability analysis using a full-scope simulator of nuclear power plants, *Proceedings of the IEEE Conference on Human Factors and Power Plants*, 2002, pp. 306–311.
134. Jung, W.S., Cho, N.Z., Semi-Markov reliability analysis of three test/repair policies for standby safety systems in a nuclear power plant, *Reliability Engineering & System Safety*, Vol. 31, No. 1, 1991, pp. 1–30.
135. Kančev, D. et al., Statistical analysis of events related to emergency diesel generators failures in the nuclear industry, *Nuclear Engineering and Design*, Vol. 273, 2014, pp. 321–331.
136. Kang, D., Jung, W.D., Yang, J.E., Improving the PSA quality in the human reliability analysis of pre-accident human errors, *Proceedings of the Annual Conference of the Canadian Nuclear Society*, 2004, pp. 189–200.

137. Kawai, K., Takishima, S., Tsuchiya, M., Uchida, M., Operator friendly man-machine system for computerized power plant automation, *IFAC Proceedings Volumes*, Vol. 17, No. 2, 1984, pp. 2641–2646.
138. Khalaquzzaman, M., Kang, H.G., Kim, M.C., Seong, P.H., A model for estimation of reactor spurious shutdown rate considering maintenance human errors in reactor protection system of nuclear power plants, *Nuclear Engineering and Design*, Vol. 240, No. 10, 2010, pp. 2963–2971.
139. Khalaquzzaman, M., Kang, H.G., Kim, M.C., Seong, P.H., Quantification of unavailability caused by random failures and maintenance human errors in nuclear power plants, *Nuclear Engineering and Design*, Vol. 240, No. 6, 2010, pp. 1606–1613.
140. Kim, I.S., Human reliability analysis in the man-machine interface design review, *Annals of Nuclear Energy*, Vol. 29, No. 11, 2001, pp. 1069–1081.
141. Kim, J., Park, J., Jung, W., Kim, J.T., Characteristics of test and maintenance human errors leading to unplanned reactor trips in nuclear power plants, *Nuclear Engineering and Design*, Vol. 239, No. 11, 2009, pp. 2530–2536.
142. Kim, S., Jung, S.H., Kim, C.H., Preventive maintenance and remote inspection of nuclear power plants using tele-robotics, *Proceedings of the International Conference on Intelligent Robots and Systems Human and Environment Friendly Robots with High Intelligence and Emotional Quotients*, 1999, pp. 603–608.
143. Kim, J., Jung, W., Park, J., AGAPE-ET: Advanced guidelines for human reliability analysis of emergency tasks, *Proceedings of the IEEE conference on Human Factors and Power Plants*, 2007, pp. 314–321.
144. Kirschenbaum, J. et al., A benchmark system for the assessment of reliability modeling methods for digital instrumentation and control systems in nuclear plants, *Proceedings of the 5th International Topical Meeting on Nuclear Plant Instrumentation Controls, and Human Machine Interface Technology*, 2006, pp. 277–283.
145. Kirwan, B., Scannali, S., Robinson, L., A case study of a human reliability assessment for an existing nuclear power plant, *Applied Ergonomics*, Vol. 27, No. 5, 1996, pp. 289–302.
146. Kirwan, B., Basra, G., Taylor-Adams, S.E., CORE-DATA: A computerized human error database for human reliability support, *Proceedings of the IEEE Conference on Human Factors and Power Plants*, 1997, pp. 7–9.
147. Knee, H.E., The maintenance personnel performance simulation (MAPPS) model: A human reliability analysis tool, *Proceedings of the International Conference on Nuclear Power Plant Aging, Availability Factor and Reliability Analysis*, 1985, pp. 77–80.
148. Koehler, T.A., Schmachtenberger, C.L., Soft approach to innovative control room design, *Instrumentation in the Power Industry*, Vol. 28, 1985, pp. 145–158.
149. Konig, N., Wetzl, R., Intelligent instrumentation and control systems for optimum power plant operation, *Engineering & Automation*, Vol. 14, No. 6, 1992, pp. 18–19.
150. Koshizuka, S., Oka, Y., Supercritical-pressure, light-water-cooled reactors for economical nuclear power plants, *Progress in Nuclear Energy*, Vol. 32, No. 3–4, 1998, pp. 547–554.
151. Koval, D.O., Floyd II, H.L., Human element factors affecting reliability and safety, *Proceedings of the IEEE Industry and Commercial Power Systems Technical Conference*, 1997, pp. 14–21.

152. Kovalevich, O.M., Verezemskii, V.G., Nuclear power plant safety and strength of equipment components with service-life extension of first-generation power-generating units, *Atomic Energy*, Vol. 90, No. 2, 2001, pp. 103–108.
153. Krasich, M., Gharakhanian, S., Qualification maintenance program plan for safety-related equipment in nuclear power generating stations, *Transactions of the American Nuclear Society*, Vol. 45, 1983, pp. 574–576.
154. Krasnodebski, J., Billinton, R., Reliability and maintainability in nuclear power generation-viewpoint of a utility, *Microelectronics and Reliability*, Vol. 15, 1976, pp. 117–118.
155. Kuppuraju, S.B., Murthy, K.S.N., Operating experience in Indian nuclear power plants of the PHWR type, *Proceedings of the International Symposium on Advances in Nuclear Power Plant Availability, Maintainability and Operation*, 1985, pp. 331–343.
156. Lee, K.N., Cho, N.Z., Semi-Markov reliability analysis of alternating systems in a nuclear power plant, *Nuclear Technology*, Vol. 98, 1992, pp. 230–241.
157. Laakso, K., Simola, K., Pulkkinen, U., Assessing the reliability of maintenance nuclear power plants, *Nuclear Europe World Scan*, Vol. 13, No. 9–10, 1993, pp. 36.
158. Lainoff, S.M., Probabilistic safety methods applied to the design of an emergency and plant information computer system nuclear power plant, *Proceedings of the International Topical Meeting on Probabilistic Safety Methods and Applications*, 1985, pp. 122–127.
159. Lang, A.W., Roth, E.M., Bladh, K., Hine, R., Using a benchmark-referenced approach for validating a power plant control room: Results of the baseline study, *Proceedings of the Human Factors and Ergonomics Society Annual Meeting*, 2002, pp. 1878–1882.
160. Laux, L., Plott, C., Using operator workload data to inform human reliability analyses, *Proceedings the IEEE Conference on Human Factors and Power Plants*, 2007, pp. 309–313.
161. Le Bot, P., Human reliability data, human error and accident models: Illustration through the Three Mile Island analysis, *Reliability Engineering and System Safety*, Vol. 83, No. 2, 2004, pp. 153–167.
162. Lee, B., The nuclear power industry's approach to human factors, *Proceedings of the IEEE Conference on Human Factors and Power Plants*, 1992, pp. 7–9.
163. Lee, J.M. et al., A communication network with high safety, maintainability, and user convenience for digital I and C systems of nuclear power plant, *Proceedings of the 8th International Conference on Emerging Technologies and Factory Automation*, 2001, pp. 353–358.
164. Lee, J.W., Oh, H.C., Lee, Y.H., Sim, B.S., Human factors research in KAERI for nuclear power plants, *Proceedings of the IEEE Annual Human Factors Meeting*, 1997, pp. 13.11–13.16.
165. Lee, J.W. et al., Human factors review guide for Korean next generation reactors, *Proceedings of the 15th International Conference on Structural Mechanics in Reactor Technology*, 1999, pp. 353–360.
166. Lee, Y.H., Kang, H.T., Chung, H.Y., Strategy for establishing integrated I&C reliability of operating nuclear power plants in Korea, *Proceedings of the International Conference on Advances in Nuclear Power Plants*, 2008, pp. 899–903.
167. Li, C. et al., Analysis on the reliability assurance program of advanced nuclear power plant operating, *Proceedings of the 10th International Conference on Reliability, Maintainability, and Safety*, 2014, pp. 1008–1011.

168. Li, X., Human reliability analysis for Guangdong nuclear power station, *Atomic Energy Science and Technology*, Vol. 27, No. 4, 1993, pp. 324–328.
169. Li, Z., Chao, H., Human factors analysis and preventive countermeasures in maintenance in nuclear power plants, *Nuclear Power Engineering*, Vol. 19, No. 1, 1998, pp. 64–68.
170. Lin, J.C., A study of online maintenance practices at U.S. nuclear plants, *Proceedings of the Annual Reliability and Maintainability Symposium*, 2002, pp. 1–5.
171. Lisboa, J., Human factors assessment of digital monitoring systems for nuclear power plants control room, *IEEE Transactions on Nuclear Science*, Vol. 39, No. 4, 1992, pp. 924–932.
172. Lisboa, J.J., Nuclear power plant availability and the role of human factors performance, *IEEE Transactions on Nuclear Science*, Vol. 37, No. 1, 1990, pp. 980–986.
173. Lois, E. et al., Capturing control room simulator data with the HERA system, *Proceedings of the IEEE Conference on Human Factors and Power Plants*, 2007, pp. 210–217.
174. Lois, E., Cooper, S.E., How do you define a human reliability analysis expert?, *Invited, Transactions of the American Nuclear Society*, Vol. 101, 2009, pp. 980–981.
175. Lu, L., Jiang, J., Probabilistic safety assessment for instrumentation and control systems in nuclear power plants: An overview, *Journal of Nuclear Science and Technology*, Vol. 41, No. 3, 2004, pp. 323–330.
176. Ma, J., Jiang, J., Applications of fault detection and diagnosis methods in nuclear power plants: A review, *Progress in Nuclear Energy*, Vol. 53, No. 3, 2011, pp. 255–266.
177. Makansi, J., Powerplant training: Ensure your team's survival in the trenches, *Power*, Vol. 139, No. 1, 1995, pp. 23–27.
178. Manicut, M., Manicut, I., Human reliability analysis-component of probabilistic safety assessment of nuclear power plant, *Proceedings of the International Symposium on Nuclear Energy*, 1993, pp. 207–211.
179. Manrique, A., Valdivia, J.C., Jimenez, A., Human factors engineering applied to nuclear power plant design, *Proceedings of the International Congress on Advances in Nuclear Power Plants*, 2008, pp. 339–346.
180. Margulies, T.S., Risk optimization: Siting of nuclear power electricity generating units, *Reliability Engineering & System Safety*, Vol. 86, No. 3, 2004, pp. 323–325.
181. Martínez-Córcoles, M., Gracia, F., Tomás, I., Peiró, J.M., Leadership and employees' perceived safety behaviors in a nuclear power plant: A structural equation model, *Safety Science*, Vol. 49, No. 8-9, 2011, pp. 1118–1129.
182. Mashin, V.A., Incidents at nuclear power plants caused by the human factor, *Power Technology and Engineering*, Vol. 46, No. 3, 2012, pp. 215–220.
183. Maurer, H.A., A note on the application of probabilistic structural reliability methodology to nuclear power plants, *Nuclear Engineering and Design*, Vol. 50, No. 2, 1978, pp. 213–215.
184. McColm, E.J., Mukherjee, P.K., Sato, J.A., Evaluation of nuclear power plant concrete to maintain continued service, *Proceedings of the Fourth International Conference on CANDU Maintenance*, 1997, pp. 336–345.
185. McElroy, A.J., IEEE project 500-reliability data manual for nuclear power generating stations, *Proceedings of the Annual Reliability and Maintainability Symposium*, 1976, pp. 277–281.

186. McKeithan, B.G., Kfoury, N.S., Increasing electric power plant productivity through maintenance management, *Proceedings of the Human Factors Society Annual Meeting, Vol. 1*, 1983, pp. 576–580.

187. Meclot, B., Planning of outages as part of the scheduling of operation for a series of nuclear power plants, *Proceedings of the International Symposium on Advances in Nuclear Power Plant Availability, Maintainability and Operation*, 1985, pp. 283–291.

188. Minner, D.E., The INEL Human Reliability Program: The first two years of experience, *Nuclear Materials Management*, Vol. 15, 1986, pp. 666–669.

189. Mirabdolbaqi, S., The role of the operator in power plant incidents, *Proceeding of the International Conference on Human Interfaces in Control Room, Cockpits and Command Centres*, 1999, pp. 276–279.

190. Misak, J. et al., Use of computer codes to improve nuclear power plant operation, *Proceedings of the International Symposium on Advances in Nuclear Power Plant Availability, Maintainability and Operation*, 1985, pp. 321–329.

191. Mishima, S., Human factors research program-long term plan in cooperation with government and private research centers, *Proceedings of the IEEE Conference on Human Factors and Power Plants*, 1988, pp. 50–54.

192. Mitenkov, F.M. et al., Improving the reliability and performance of a nuclear power plant by using computers in control systems, *Soviet Atomic Energy*, Vol. 66, No. 6, 1989, pp. 479–483.

193. Miyazaki, T., Development of a new cause classification method considering plant aging and human errors for adverse events occurred in nuclear power plants and some results of its application, *Transactions of the Atomic Energy Society of Japan*, Vol. 6, No. 4, 2007, pp. 434–443.

194. Mookerjee, G., Mandal, J., Modern control center design—The influence of operator interface on control room instrumentation and control devices, considering the philosophy of human-factor engineering principles, *Proceedings of the American Power Conference*, 1998, pp. 1009–1014.

195. Mookerjee, G., Modern control center design—the influence of operator interface on control room instrumentation and control devices, *Considering the Philosophy of Human—Factors Engineering Principles, Proceedings of the American Power Conference on Reliability and Economy*, 1998, pp. 1009–1014.

196. Morgan, T.A., A reliability-based bounding analysis methodology for seismic isolated nuclear power plants: Safety, reliability, risk and life-cycle performance of structures and infrastructures, *Proceedings of the 11th International Conference on Structural Safety and Reliability*, 2013, pp. 4245–4250.

197. Morgenstern, M.H., Maintenance management systems: A human factors issue, *Proceedings of the IEEE Conference on Human Factors and Power Plants*, 1988, pp. 390–393.

198. Mosneron-Dupin, F., Saliou, G., Lars, R., Probabilistic human reliability analysis: The lessons derived for plant operation at Electric de France, *Reliability Engineering and System Safety*, Vol. 58, No. 3, 1997, pp. 249–274.

199. Mosneron-Dupin, F., Villemeur, A., Moroni, J.M., Paluel nuclear power plant PSA: Methodology for assessing human reliability, *Proceedings of the 7th International Conference on Reliability and Maintainability*, 1990, pp. 584–590.

200. Mrowca, B., Dube, D.A., Appignani, P.L., Clark, J.B., Design insights resulting from the comparison of U.S. nuclear power plant component Birnbaum importance measures, *Proceedings of the 8th International Conference on Probabilistic Safety Assessment and Management*, 2006, pp. 260–265.

201. Mukhopadhyay, S., Chaudhuri, S., A feature-based approach to monitor motor operated valves used in nuclear power plants, *IEEE Transactions on Nuclear Science*, Vol. 42, No. 6, 1995, pp. 2209–2220.
202. Mulle, G.R., Dick, R., The role of human factors in planning for nuclear power plant decommissioning, *Proceedings of the IEEE Conference on Human Factors and Nuclear Safety*, 1985, pp. 257–262.
203. Muzzammil, M.H.S.M., Ali, E.A.M., Safety improvement of nuclear power reactor using soft computing techniques, *Proceedings of the International Conference on Energy Efficient Technologies for Sustainability*, 2013, pp. 949–954.
204. Nagata, T., Sugiyama, K.I., Performance evaluation of Japanese nuclear power plant based on open data and information, *Proceedings of the 17th International Conference on Nuclear Engineering*, 2009, pp. 217–224.
205. Naser, J., Addressing digital control room human factors technical and regulatory issues, *Proceedings of the American Nuclear Society International Meeting on Nuclear Plant Instrumentation, control and Human-Machine Interface Technology*, 2009, pp. 6–10.
206. Naser, J. et al., Nuclear power plant control room modernization planning, process, human factors engineering, and licensing guidance, *Proceedings of the Annual Joint ISA Power Industry Division and EPRI Controls and Instrumentation Conference*, 2004, pp. 219–227.
207. Naumov, V.I., Human factors and supporting measures for nuclear power plant operators, *Atomnaya Energiya*, Vol. 74, No. 4, 1993, pp. 344–348.
208. Nelson, W.R., Integrated design environment for human performance and human reliability analysis, *Proceedings of the IEEE conference on Human Factors and Power Plants*, 1997, pp. 8.7–8.11.
209. O'Brien, J.N., Lukas, W.J., A strategy for examining human reliability aspects of plant security, *Proceedings of the IEEE Conference on Human Factors and Power Plants*, 1998, pp. 465–470.
210. O'Hara, J.M. et al., Updating the NRC's guidance for human factors engineering reviews, *Proceedings of the IEEE Conference on Human Factors and Power Plants*, 2002, pp. 422–427.
211. O'Hara, J.M., Hall, R.E., Advanced control rooms and crew performance issues: Implications for human reliability, *IEEE Transactions on Nuclear Science*, Vol. 39, No. 4, 1992, pp. 919–923.
212. Ohga, Y., Nagaoka, Y., Suzuki, S., Ito, T., Natural language interface for fault diagnosis system of nuclear power plant control systems, *Journal of Nuclear Science and Technology*, Vol. 27, No. 9, 1990, pp. 790–801.
213. Onodera, K., Miki, M., Nukada, K., Nakamura, H., Reliability management of nuclear power plant, *Proceedings of the Annual Reliability and Maintainability Symposium*, 1982, pp. 151–156.
214. Orendi, R.G. et al., Human factors consideration in emergency procedure implementation, *Proceedings of the IEEE Conference on Human Factors and Power Plants*, 1988, pp. 214–221.
215. Oxstarnd, J., Boring, R.L., Human reliability for design applications at a Swedish nuclear power plant, *Proceedings of the International Symposium on Resilient Control Systems*, 2009, pp. 5–10.
216. Pack, R.W., Parris, H.L., Human factors engineering design guidelines for maintainability, *Proceedings of the Inter-RAM Conference for the Electric Power Industry*, 1985, pp. 11–14.

217. Paradies, M., Positive vs. negative enforcement: Which promotes high reliability human performance, *Proceedings of the IEEE Conference on Human Factors and Power Plants*, 2007, pp. 185–188.

218. Park, J.W., Jung, W., Comparing cultural profiles of MCR operators with those of non-MCR operators working in domestic nuclear power plants, *Reliability Engineering & System Safety*, Vol. 133, 2015, pp. 146–156.

219. Park, J.Y., Jerng, D.W., A method to monitor the reliability of in-house power supply systems in nuclear power plants based on probabilistic assessment, *Transactions of the Korean Institute of Electrical Engineers*, Vol. 58, No. 3, 2009, pp. 444–449.

220. Parsons, S.O., Seminara, J.L., O'Hara, J.M., Power system human factors/ergonomics activities in United States, *Proceedings of the XIV Triennial Congress of the International Ergonomics Association and 44th Annual Meeting of Human Factors and Ergonomic Association*, 2000, pp. 807–810.

221. Penington, J., Shakeri, S., A human factors approach to effective maintenance, *Proceedings of the Canadian Nuclear Society Conference*, 2006, pp. 1–11.

222. Pereguda, A.I., Chekhovich, V.E., Sukhetskii, A.K., Reliability estimation for important nuclear power station safety equipment, *Atomic Energy*, Vol. 73, No. 4, 1992, pp. 784–788.

223. Pereguda, A.I., Petrenko, A.A., Ensuring required reliability for nuclear reactor protection systems, *Soviet Atomic Energy*, Vol. 67, No. 6, 1989, pp. 859–863.

224. Pesme, H., Le Bot, P., Meyer, P., Little stories to explain human reliability assessment: A practical approach of MERMOS method, *Proceedings of the IEEE Conference on Human Factors and Power Plants*, 2007, pp. 284–287.

225. Peterson, G.R., Industry initiative to improve maintenance at nuclear power plants, *Proceedings of the Annual Reliability and Maintainability Symposium*, 1989, pp. 415–417.

226. Phan, H.K., Pham, H., Reliability analysis of emergency telecommunication systems in nuclear power plants, *Proceedings of the Annual Reliability and Maintainability Symposium*, 1994, pp. 40–45.

227. Pine, S.M., Schulz, K.A., Applying human engineering processes to control room design, *Power Engineering* Vol. 87, No. 1, 1983, pp. 38–46.

228. Presensky, J.J., Human factors and power plants: Foreword, *Proceedings of the IEEE Conference on Human Factors and Power Plants*, 2002, pp. 3–4.

229. Purba, J.H., A fuzzy-based reliability approach to evaluate basic events of fault tree analysis for nuclear power plant probabilistic safety assessment, *Annals of Nuclear Energy*, Vol. 70, 2014, pp. 21–29.

230. Purba, J.H., Fuzzy probability on reliability study of nuclear power plant probabilistic safety assessment: A review, *Progress in Nuclear Energy*, Vol. 76, 2014, pp. 73–80.

231. Pyy, P., Laakso, K., Reiman, L., A study on human errors related to NPP maintenance activities, *Proceedings of the IEEE Conference on Human Factors and Power Plants*, 1997, pp. 12.23–12.28.

232. Racek, J., Protection against corrosion of the primary circuit of nuclear power plant with gas-cooled reactor, *Proceedings of the 13th International Scientific Conference on Electric Power Engineering*, 2012, pp. 1339–1342.

233. Raghavan, R., Simon, B.H., Advanced plant maintenance and surveillance system for the nuclear power plants of the next century, *Proceedings of the 2nd ASME JSME Nuclear Engineering Joint Conference*, 1993, pp. 693–697.

234. Ravindra, M.K., System reliability considerations in probabilistic risk assessment of nuclear power plants, *Structural Safety*, Vol. 7, No. 2–4, 1990, pp. 269–280.

235. Reiman, T., Oedewald, P., Assessing the maintenance unit of a nuclear power plant—identifying the cultural conceptions concerning the maintenance work and the maintenance organization, *Safety Science*, Vol. 44, No. 9, 2006, pp. 821–850.

236. Reiman, T., Oedewald, P., Organizational factors and safe human performance—Work physiological model, *Proceedings of the IEEE Conference on Human Factors and Power Plants*, 2007, pp. 140–144.

237. Robbins, M.C., Eames, G.F., Mayell, J.R., Nuclear safety chains, *IEEE Proceedings*, Vol. 128, No. 2, 1981, pp. 100–107.

238. Robinson, D.G., Performance and reliability monitoring of advanced reactors, *Proceedings of the 5th International Topical Meeting on Nuclear Plant Instrumentation Controls, and Human Machine Interface Technology*, 2006, pp. 807–813.

239. Rosen, Y.G., Nyh, L.N., Availability study of for SMARK 3 nuclear power plant, *Proceedings of the Annual Reliability and Maintainability Symposium*, 1980, pp. 70–75.

240. Ruf, R., Hummeler, A., Steam generator replacement at the Obrigheim Nuclear Power Plant: Nuclear power plant outage experience, *Proceedings of the International Symposium on Advances in Nuclear Power Plant Availability, Maintainability and Operation*, 1984, pp. 299–311.

241. Runow, P., Maurer, H.A., Current state of acoustic emission as an aid to the structural integrity assessment of nuclear power plants, *Proceedings of the International Symposium on Advances in Nuclear Power Plant Availability, Maintainability and Operation*, 1985, pp. 175–187.

242. Ryan, T.G., A task analysis-linked approach for integrating the human factor in reliability assessments of nuclear power plants, *Reliability Engineering and System Safety*, Vol. 22, No. 1–4, 1988, pp. 219–234.

243. Sa, K.K. et al., An investigation on unintended reactor trip events in terms of human error hazards of Korean nuclear power plants, *Annals of Nuclear Energy*, Vol. 65, 2014, pp. 223–231.

244. Saldanha, P.L.C., Ferro, N.J., Frutoso e Melo, P.F., Marques, F.F., Reliability attributes and the maintenance rule for nuclear power plant safety systems, *Chemical Engineering Transactions*, Vol. 33, 2013, pp. 883–888.

245. Salge, M, Milling, P.M., Who is to blame, the operator or the designer? Two stages of human failure in the Chernobyl accident, *System Dynamics Review*, Vol. 22, No. 2, 2006, pp. 89–112.

246. Sanwarwalla, M.H., Reliability of onsite emergency power for the new generation and advanced nuclear power plants, *Proceedings of the ASME Pressure Vessels and Piping Conference*, 2010, pp. 25–29.

247. Schroeder, L.R., Gaddy, C.D., Beare, A.N., New control technologies require good human factors engineering, *Power Engineering*, Vol. 93, No. 11, 1989, pp. 29–32.

248. Sedlak, J., Reliability assessment of diversity in digital I&C systems at nuclear power plants, *Proceedings of the 7th International Topical Meeting on Nuclear Plant Instrumentation, Control, and Human-Machine Interface Technologies*, 2010, pp. 1272–1281.

249. Seminara, J.L., Parsons, S.O., Human-factors engineering and power plant maintenance, *Maintenance Management International*, Vol. 6, No. 1, 1980, pp. 33–71.

250. Seminara, J.L., Parsons, S.O., Survey of control-room design practices with respect to human factors engineering, *Nuclear Safety*, Vol. 21, No. 5, 1980, pp. 603–617.
251. Seminara, J.L., Parsons, S.O., Nuclear power plant maintainability, *Applied Ergonomics*, Vol. 13, No. 3, 1982, pp. 177–189.
252. Shrikhande, S.V. et al., Hardware reliability prediction of computer based safety systems of Indian nuclear plants, *Proceedings of the 2nd International Conference on Reliability, Safety and Hazard, Risk-Based Technology and Physics-of-Failure Methods*, 2010, pp. 127–132.
253. Shviryaev, Y.V., Derzhinskii, F.E., Bukrinskii, A.M., Requirements for reliability of nuclear power plant safety systems, *Soviet Power Engineering*, Vol. 10, No. 3, 1981, pp. 323–330.
254. Skjolde, B.R., White, D., The IAEA power reactor information system and its possible use for improving plant performance, *Maintenance and Operation, Proceedings of the International Symposium on Advances in Nuclear Power Plant Availability, Maintainability and Operation*, 1985, pp. 17–28.
255. Smith, D.J., Integrated control systems: The next step, *Power Engineering*, Vol. 95, No. 9, 1991, pp. 17–21.
256. So, A.T.P., Chan, W.L., A computer-vision based power plant monitoring system, *Proceedings of the International Conference on Advances in Power System Control, Operation, and Management*, 1991, pp. 335–340.
257. Soon, H.C. et al., Development of an advanced human-machine interface for next generation nuclear power plants, *Reliability Engineering & System Safety*, Vol. 64, No. 1, 1999, pp. 109–126.
258. Spurgin, A.J., Lydell, B.O.Y., Critique of current human reliability analysis methods, *Proceedings of the IEEE Conference on Human Factors and Power Plants*, 2002, pp. 312–318.
259. Srinivas, G. et al., Hardware reliability assessment of safety related and safety critical systems in nuclear power plants, *Proceedings of the 2nd International Conference on Reliability, Safety and Hazard—Risk-Based Technologies and Physics of-Failure Methods*, 2010, pp. 448–454.
260. Staples, L., McRobbie, H., Design changes and human factors, *Nuclear Plant Journal*, Vol. 22, No. 1, 2004, pp. 41–45.
261. Sternheim, E., Moser, T.D., System design considerations for high powered testing with real time computers, *Proceedings of the Power Industry Computer Applications Conference*, 1977, pp. 407–410.
262. Stojka, T., Use of expert judgements for assessment of performance shaping factors in human reliability analysis, *Proceedings of the American Nuclear Society International Topical Meeting on Probabilistic Safety Analysis*, 2005, pp. 619–623.
263. Strater, O., Bubb, H., Assessment of human reliability based on evaluation of plant experience: Requirements and implementation, *Reliability Engineering & System Safety*, Vol. 63, No. 2, 1999, pp. 199–219.
264. Strod, A.A., The importance of human factors in performing maintenance tasks inside the containment of a nuclear power plant, *Proceedings of the Annual Engineering Conference on Reliability for the Electric Power Industry*, 1980, pp. 82–87.
265. Stultz, S.W., Schroeder, L.R., Incorporating human engineering principles in distributed controls upgrades, *Instrumentation, Control, and Automation in the Power Industry*, Vol. 31, 1988, pp. 135–141.

266. Subudhi, M., Taylor, J., Improving motor reliability in nuclear power plants, *Proceedings of the U.S. Nuclear Regulatory Commission Fifteenth Water Reactor Safety Information Meeting*, 1987, pp. 109–113.

267. Sung, C., Chung, H., Cho, S., A methodology for qualitative reliability evaluation of I&C systems in nuclear power plants, *Proceedings of the American Nuclear Society International Congress on Advances in Nuclear Power Plants*, 2005, pp. 2016–2024.

268. Swain, A.D., Guttmann, H.E., Human reliability analysis applied to nuclear power, *Proceedings of the Annual Reliability and Maintainability Symposium*, 1974, pp. 116–119.

269. Swaton, E., Neboyan, V., Lederman, L., Human factors in the operation of nuclear power plants, *International Atomic Energy Bulletin*, Vol. 29, No. 4, 1987, pp. 27–30.

270. Sword, J., How do you know what you don't know? Detecting "hidden trends" in performance, *Proceedings of the IEEE Conference on Human Factors and Power Plants*, 2007, pp. 38–41.

271. Szoch Jr., R.L., Brown, E.M., Wilkinson, C.D., Strategic alliance to address safety system obsolescence and maintainability, *Proceedings of the IEEE Nuclear Science Symposium & Medical Imaging Conference*, 1995, pp. 1073–1076.

272. Taniguchi, T., RD & D for improving reliability of Japanese nuclear power plants, *Nuclear Engineering and Design*, Vol. 87, 1985, pp. 139–152.

273. Tarasenko, V.M., Nuclear power plant compliance with the principle of power-system independence from safety systems operation, *Atomic Energy*, Vol. 94, No. 2, 2003, pp. 76–81.

274. Tashjian, B.M., Planning for nuclear plant reliability, availability, and maintainability, *Proceedings of the 5th Annual National Conference on Nuclear Power*, 1978, pp. A1–14.

275. Tasset, D., Labarthe, J.P., The impact on safety of computerized control room in nuclear power plants: The French experience on human factors with N4 series, *Proceedings of the IEA/HFES Congress*, 2000, pp. 815–818.

276. Tennant, D.V., Human factors considerations in power plant control room design, *Instrumentation in Power Industry*, Vol. 29, 1986, pp. 29–36.

277. Thomas, S., Economic and safety pressures on nuclear power: A comparison of Russia and Ukraine since the break-up of the soviet union, *Energy Policy*, Vol. 27, No. 13, 1999, pp. 745–767.

278. Thunberg, A., Osvalder, A.L., What constitutes a well-designed alarm system?, *Proceedings of the IEEE Conference on Human Factors and Power Plants*, 2007, pp. 85–91.

279. Tinell, T., Kallio, H., Raumolin, H., Advances in the continuous improvement in quality and safety at Loviisa Nuclear Power Plant, *Proceedings of the International Symposium on Advances in the Operational Safety of Nuclear Power Plants*, 1996, pp. 131–141.

280. Tonkinson, T.S., SAI's behaviour-based root cause analysis, *Proceedings of the IEEE Conference on Human Factors and Power Plants*, 2007, pp. 288–290.

281. Trager, E.A., Human errors in events involving wrong unit or wrong train, *Nuclear Safety*, Vol. 25, No. 5, 1983, pp. 697–703.

282. Tran, T.Q., Boring, R.L., Joe, J.C., Griffith, C.D., Extracting and converting quantitative data into human error probabilities, *Proceedings of the IEEE Conference on Human Factors and Power Plants*, 2007, pp. 164–169.

283. Ujita, H., Human error classification and analysis in nuclear power plants, *Journal of Nuclear Science and Technology*, Vol. 22, No. 6, 1985, pp. 496–498.

284. Upadhyaya, B.R., Perillo, S.R.P., Xu, X., Li, F., Advanced control design, optimal sensor placement, and technology demonstration for small and medium nuclear power reactors, *Proceedings of the 17th International Conference on Nuclear Engineering*, 2009, pp. 763–773.

285. Van Cott, H.P., The application of human factors by DOE nuclear facilities, *Proceedings of the Human Factors Society Annual Meeting*, 1984, pp. 138–139.

286. Varma, V., Maintenance training reduces human errors, *Power Engineering*, Vol. 100, No. 8, 1996, pp. 44–47.

287. Varnicar, J.J., Use of mockups for control panel enhancements, *Proceedings of the IEEE Conference on Human Factors and Power Plants*, 1985, pp. 41–44.

288. Vaurio, J.K., Human factors, human reliability and risk assessment in license renewal of a nuclear power plant, *Reliability Engineering & System Safety*, Vol. 94, No. 11, 2009, pp. 1818–1826.

289. Verlinden, S., Deconinck, G., Coupe, B., Hybrid reliability model for nuclear reactor safety system, *Reliability Engineering & System Safety*, Vol. 101, 2012, pp. 35–47.

290. Vesely, W.E., Goldberg, F.F., Time dependent unavailability analysis of nuclear safety systems, *IEEE Transactions on Reliability*, Vol. 26, No. 4, 1977, pp. 257–260.

291. Volkanovski, A., Gjorgiev, B., Renewable sources impact on power system reliability and nuclear safety, *Proceedings of the European Safety and Reliability Conference*, 2015, pp 57–63.

292. Volkanovski, A., On-site power system reliability of a nuclear power plant after the earthquake, *Kerntechnik*, Vol. 78, No. 2, 2013, pp. 99–112.

293. Voss, T.J., An overview of IEEE human factors standard activities, *Proceedings of the American Nuclear Society International Topical Meeting on Nuclear Plant Instrumentation, Control and Human Machine Interface Technology*, 2004, pp. 875–880.

294. Wei, L., He, X.H., Zhao, B.Q., Research on the relationship between nuclear power plant operator's reliability and human quality, *Atomic Energy Science and Technology*, Vol. 38, No. 4, 2004, pp. 312–316.

295. Wells, D.J., Hazardous area robotics for nuclear systems maintenance: A challenge in reliability, *Proceedings of the Portland International Conference: Management on Engineering Technology*, 1992, pp. 371–373.

296. Whaley, A.M. et al., Lessons learned from dependency usage in HERA: Implications for THERP-related HRA methods, *Proceedings of the IEEE Conference on Human Factors and Power Plants*, 2007, pp. 322–327.

297. Widrig, R.D., Human factors: A major issue in plant aging, *Proceedings of the International Conference on Nuclear Power Plant Aging, Availability Factor, and Reliability Analysis*, 1985, pp. 65–68.

298. Williams, J.C., A data-based method for assessing and reducing human error to improve operational performance, *Proceedings of the IEEE Conference on Human Factors and Power Plants*, 1988, pp. 436–450.

299. Wise, M.J., Johnson, R.D., Nuclear plant reliability data program, *Proceedings of the Annual Reliability and Maintainability Symposium*, 1975, pp. 143–148.

300. Woo, T.H., Lee, U.C., The statistical analysis of the passive system reliability in the nuclear power plants (NPPs), *Progress in Nuclear Energy*, Vol. 52, No. 5, 2010, pp. 456–61.

301. Woo, T.H., Lee, U.C., Safety assessment for the passive system of the nuclear power plants (NPPs) using safety margin estimation, *Energy*, Vol. 35, No. 4, 2010, pp. 1799–1804.
302. Woo, T.H., Lee, U.C., Passive system reliability in the nuclear power plants (NPPs) using statistical modeling, *Nuclear Engineering and Design*, Vol. 239, No. 12, 2009, pp. 3014–3020.
303. Ding, C.G., Lin, H.R., Wu, C.H., Jane, T.D., Using LGM analysis to identify hidden contributors to risk in the operation of a nuclear power plant, *Safety Science*, Vol. 75, 2015, pp. 64–71.
304. Wu, T.M., Lee, J.Y., A large scale implementation of human factors engineering for the Lungmen Nuclear Power Plant, *Proceedings of the 4th American Nuclear Society International Topical Meeting on Nuclear Plant Instrumentation, Controls and Human-Machine Interface Technology*, 2004, pp. 1–10.
305. Xiong, G. et al., Parallel system method to improve safety and reliability of nuclear power plants, *Proceedings of the 9th World Congress on Intelligent Control and Automation*, 2011, pp. 237–242.
306. Yamaguchi, M., Ohshita, Y., Arafune, K., Importance of photovoltaics learned from the Fukushima Nuclear Power Plant accident in Japan, *Proceedings of the 28th European Photovoltaic Solar Energy Conference and Exhibition*, 2013, pp. 4623–4628.
307. Yong S.S. et al., Plant information processing system for Korean future nuclear power plants, *Proceedings of the American Nuclear Society International Topical Meeting on Nuclear Plant Instrumentation, Control and Human-Machine Interface Technologies*, 1996, pp. 1553–1560.
308. Yoon, W.C., Lee, Y.H., Kim, Y.S., A model-based and computer-aided approach to analysis of human errors in nuclear power plants, *Reliability Engineering & System Safety*, Vol. 51, No. 1, 1996, pp. 43–52.
309. Young, R., UNIRAM modeling of nuclear power plants to support availability improvement, *Proceedings of the 13th INTER-RAM: International Reliability, Availability, Maintainability Conference for the Electric Power Industry*, 1985, pp. 7–9.
310. Yu, K., Gutierrez, R., Zizzo, D., Human factors verification of the advanced nuclear plant control room design, *Proceedings of the Annual Joint ISA POWID/EPRI Controls and Instrumentation Conference*, 2008, pp. 252–263.
311. Yu, Y., Jiejuan, T., Ruichang, Z., Aling, Z., Reliability analysis for continuous operation system in nuclear power plant, *Proceedings of the 8th International Conference on Reliability, Maintainability, and Safety*, 2009, pp. 171–173.
312. Zhang, J. et al., Parallel system method to improve safety and reliability of nuclear power plants, *Proceedings of the 9th World Congress on Intelligent Control and Automation*, 2011, pp. 237–242.
313. Zhang, Q., Mclellan, B.C., Review of Japan's power generation scenarios in light of the Fukushima nuclear accident, *International Journal of Energy Research*, Vol. 38, No. 5, 2014, pp. 539–550.
314. Zhang, Z., Hu, L., Performance assessment for the water level control system in steam generator of the nuclear power plant, *Proceedings of the IEEE International Conference on Multimedia and Expo*, 2011, pp. 5842–5847.
315. Zhao, T. et al., Based on multi-agent optimize the operation of the power plant unit of assessment management system, *IET Seminar Digest*, Vol. 1, No. 3, 2008, pp. 214–219.

316. Zhen, L. et al., Reliability research of power-supply system in nuclear power plant based on Markov theory and Bayes network, *Proceedings of the International Conference on Nuclear Engineering*, 2010, pp. 120–125.

317. Zhuravyov, G.E., Sakov, B.A., Ergonomic and psychological provisions of a power plant, *Proceedings of the Triennial Congress of the International Ergonomics Association and the Annual Meeting of the Human Factors and Ergonomics Association*, 2000, pp. 831–834.

318. Zio, E., Maio, F.D., Processing dynamic scenarios from a reliability analysis of a nuclear power plant digital instrumentation and control system, *Annals of Nuclear Energy*, Vol. 36, No. 9, 2009, pp. 1386–1399.

319. Zubair, M. et al., A computer based living probabilistic safety assessment (LPSA) method for nuclear power plants, *Nuclear Engineering and Design*, Vol. 265, 2013, pp. 765–771.

320. Zubair, M., Zhang, Z., Reliability data update method for emergency diesel generator of Daya Bay Nuclear Power Plant, *Annals of Nuclear Energy*, Vol. 38, No. 11, 2011, pp. 2575–2580.

Index